CANINES IN CERVANTES AND VELÁZQUEZ

The study of the creation of canine breeds in early modern Europe, especially Spain, illustrates the different constructs against which notions of human identity were forged. This book is the first comprehensive history of early modern Spanish dogs and it evaluates how two of Spain's most celebrated and canonical cultural figures of this period, the artist Diego Velázquez and the author Miguel de Cervantes, radically question humankind's sixteenth-century anthropocentric self-fashioning. In general, this study illuminates how Animal Studies can offer new perspectives to understanding Hispanism, giving readers a fresh approach to the historical, literary, and artistic complexity of early modern Spain.

New Hispanisms: Cultural and Literary Studies

Series editor: Anne J. Cruz

"New Hispanisms: Literary and Cultural Studies" presents innovative studies that seek to understand how the cultural production of the Hispanic world is generated, disseminated, and consumed. Ranging from the Spanish Middle Ages to modern Spain and Latin America, this series offers a forum for various critical and disciplinary approaches to cultural texts, including literature and other artifacts of Hispanic culture. Queries and proposals for single author volumes and collections of original essays are welcome.

Reading Inebriation in Early Colonial Peru
Mónica P. Morales

Gendering the Crown in the Spanish Baroque Comedia
María Cristina Quintero

Canines in Cervantes and Velázquez

An Animal Studies Reading of Early Modern Spain

JOHN BEUSTERIEN

Routledge
Taylor & Francis Group
LONDON AND NEW YORK

First published 2013 by Ashgate Publishing

Published 2016 by Routledge
2 Park Square, Milton Park, Abingdon, Oxfordshire OX14 4RN
711 Third Avenue, New York, NY 10017, USA

First issued in paperback 2016

Routledge is an imprint of the Taylor & Francis Group, an informa business

British Library Cataloguing in Publication Data
Beusterien, John.
Canines in Cervantes and Velazquez: An Animal Studies Reading of Early Modern Spain. – (New hispanisms)
1. Velazquez, Diego, 1599–1660 – Themes, motives. 2. Cervantes Saavedra, Miguel de, 1547–1616 – Themes motives. 3. Dogs in art. 4. Dogs in literature. 5. Spain – Civilization – 1516–1700
I. Title II. Series
704.9'4329772'0946-dc23

The Library of Congress has cataloged the printed edition as follows:
Beusterien, John.
Canines in Cervantes and Velazquez: An Animal Studies Reading of Early Modern Spain / by John Beusterien.
pages cm. — (New Hispanisms: Cultural and Literary Studies)
Includes index.
ISBN 978-1-4094-5713-8 (hardcover: alk. paper)
1. Spanish literature—Classical period, 1500–1700—History and criticism. 2. Cervantes Saavedra, Miguel de, 1547–1616—Criticism and interpretation. 3. Velazquez, Diego, 1599–1660—Criticism and interpretation. 4. Dogs in literature. 5. Dogs in art. 6. Dogs—Spain. I. Title.
PQ6066.B47 2013
863'.3—dc23

2012040944

ISBN 13: 978-1-138-27190-6 (pbk)
ISBN 13: 978-1-4094-5713-8 (hbk)

In memory of my father, who first saw Bruna smile.

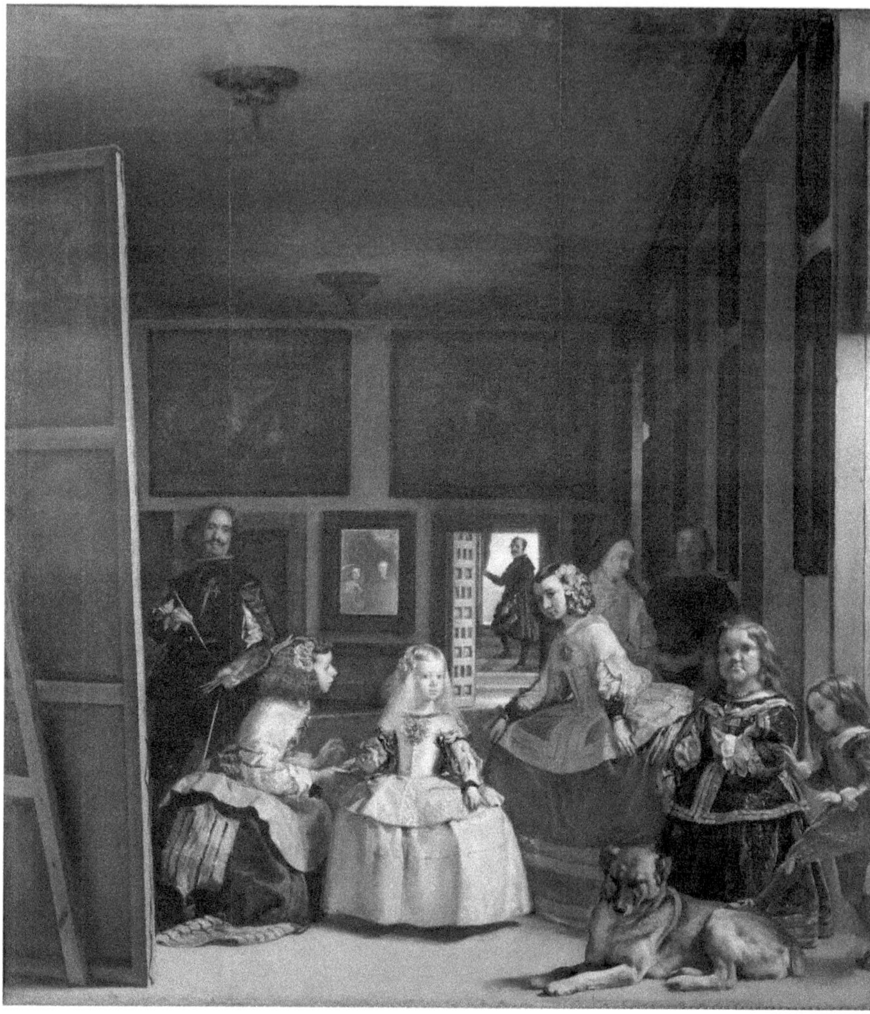

Diego Velázquez, *Las meninas*, 1656, Madrid, Prado.

Contents

List of Figures

Acknowledgements

Thanks to Texas Tech University, the Program for Cultural Cooperation between Spain's Ministry of Culture and United States' Universities, and the Charles Redd Center for Western Studies at Brigham Young University for providing research support at the various stages of the writing of the manuscript. Thanks also to Texas Tech University for the financial support for the reproduction of images. Thanks to *Cervantes: Bulletin of the Cervantes Society of America* for giving me permission to use previously published material (from the article "'El nombre de podenco': The Dog as Book in the Prologue of Part II of *Don Quijote*"). Thanks to Erika Gaffney for her keen editorial eye and help in getting the manuscript ready for publication. Finally, a special thanks to Anne Cruz for her perceptive comments, confidence, and support.

Introduction

Gaspar Pérez de Villagrá published *The History of New Mexico* (*Historia de la Nueva México*) in 1610 in Spain. Describing events that took place in New Mexico, the book is the first epic poem about a future U.S. state, since the region described in the poem would become part of the 47th state over three hundred years later (in 1912). The poem recounts the early years of the Oñate expedition that had set out in 1598 to create a Spanish settlement in New Mexico. Pérez de Villagrá, the poet, was one of Oñate's captains and his work is an example of the revival of the classic epic, a popular genre in early modern Spain. In standard epics from the period, the Spanish dog in the Americas exemplified the ideal Spanish warrior. Companion to sixteenth-century conquistadors, the Spanish dog was a mighty military fighting machine and, when it died, it was mourned as a heroic fallen soldier.

The dog described in the *History of New Mexico*, however, is quite different from the conventional heroic dogs described in other Spanish epics. The dog has no name and it is not described as a soldier. When the dog dies, Villagrá does not sing of its heroic feats. Just as Oñate's mission would ultimately end in failure (the Pueblo peoples forced the Spaniards to leave in 1680), so Pérez de Villagrá's relationship with his dog ends up in ruins. In a pitiful, heart-wrenching scene, he tries to kill the dog for food, and, afterwards, the dog that he has tried to kill returns to him, not only licking the blood of its own wounds, but the blood on the master's hands who has just delivered the fatal blow. Pérez de Villagrá finally leaves the dog's dead body somewhere on the barren, uncharted land that would become New Mexico.[1]

[1] The complete text reads: "I had a dog, then, that which at my side / Had been for a long time and which did watch / What times it might be that I slept at night; / And since hunger now threatened me / So as to take away my life, I then / Determined to kill him and gave two wounds, / Both mortal, at which the poor wretch / Fled from me away some distance. / I called him angrily, and forgetful / Of the shameful and inconsiderate deed, / Whining in friendly sort and crouching down, / My badly wounded friend came back to me, / Licking the blood that he poured out. / And, though in pain and wounded sore, / To please me in something if he but could, / He also licked my hands, till they / Were stained and well bathed with his blood. / ... With such crass ignorance as not to see / That I did lack a fire to cook his flesh, / I lowered my sad eyes and, beginning, / Repenting of my deed, to caress him, / He lay dead at my feet. And with this shock, / Leaving him stretched out in his blood, / I gave a bitter gulp and went to find / That stroke of fortune, which might end / The miserable life I led" ("Llevaba, pues, un perro que a mi lado / Anduvo mucho tiempo y que velaba / Cuando de noche acaso me dormía; / Y porque ya el hambre me afligía / Por suerte que la vida me acababa / Determiné matarle y dos heridas / Le di mortales con que luego el pobre / De mí se fue apartando un largo trecho. / Llaméle con enojo y olvidado / Del vergonzoso hecho, inadvertido, / Gimiendo mansamente y agachado, / A mí volvió el amigo mal herido, / Lamiéndose la sangre que vertía. / Y así, con desconsuelo y lastimado, / Por agradarme en algo si pudiese, / Lamió también mis manos, que teñidas / Me puso, de

The continental philosopher Jacques Derrida provides a term and some clues on how to read Pérez de Villagrá's story about his dog. In *The Gift of Death,* Jacques Derrida introduces the idea of the "other other" for the animal in questioning the meaning of the human in the Abraham-Isaac sacrifice story. Killing one's own child for a greater purpose is, as Derrida puts it with respect to Abraham and Isaac, monstrous, outrageous and barely conceivable. But, as Derrida writes, we are unwittingly doing just that, killing those to whom we owe absolute fidelity day and night at every instant (67-8). In the logic of Immanuel Levinas, a singular human other contains the infinite and unknowable and the destruction of that human life, especially one's child, would be the ultimate abomination. Directly engaging Levinas and his notion of the human other, Derrida writes, following his comments on Abraham and Isaac, that animals are "even more other others than my fellows" (69). With the "other other," Derrida separates the animal from the human other.[2] He thereby does not provide support for arguments, as some have argued, like the execution of humans in the holocaust is equal with the execution of animals in slaughterhouses.[3] Instead, Derrida's "other other" looks to an ethics based on the infinite and unknowable number of others to whom responsibility is due.

In contrast to the Spanish hero dogs (and other famed classical and humanist dogs) that had mausoleums and memorials in early modern Europe, the body of Pérez de Villagrá's dog is left abandoned on the outer reaches of Spain's sixteenth-century northern frontier in America. Pérez de Villagrá wishes to evoke sympathy for the suffering animal. Some readers of the scene might be inclined to interpret the body of the dog as a metaphor for the life of Pérez de Villagrá and the struggling Spanish imperial enterprise on the northern New Spain frontier. But, although the dog has suffered, its suffering is not the same as that of a person. The body of the abandoned dog deserves an empathy not based on the Levinas ethical logic of the uniqueness of the human face. The dog does not have a human face, or the name of a human. It is not a scapegoat for the human. Pérez de Villagrá's love for his dog is an *amor perro*, an expression that suggests a dog, but in which *perro* does not literally mean "dog," but is an adjective meaning "messed-up" or "ruined" and modifying "love."[4] The "messed-up love" for his dog exists outside of sacrificial

su sangre bien bañadas./ ...Con tan crasa ignorancia que no vía / Que fuego para asarlo me faltaba, / Bajé los ojos tristes y volviendo, / Del hecho arrepentido, a acariciarlo, / Muerto quedó a mis pies. Con cuyo susto, / Dejándolo tendido y desangrado, / Pasé aquel trago amargo y fui siguiendo / El golpe de fortuna que acababa / La miserable vida que vivía" [236–48]).

2 Donna Haraway also defines the animal as a unique other. See her study on the dog and "significant otherness" (*The Companion*).

3 For a brief synopsis on this explosively polemical conclusion about Derrida, see Wolfe's comments in Cavalieri 126.

4 The negative connotations of *perro* as adjective can be found in cues from common expressions that indicate the existence of a negative attitude towards dogs in Spanish-speaking cultures. The expression *día de perros* is common among most Spanish-speaking peoples and it means a "dog day," that is, a really bad day. To have a *perra* can also mean

logic based on traditional definitions of the human. The dog is both friend and enemy and is neither. The dog is friend and "human" in the sense that it suffers and demands sympathy. But it is other and "animal" in the sense that it is *not* an animal that is sacrificed and mourned like the soldier dogs that populate other epics from the period. It does not perform war feats, its flesh is considered for food, and, nameless, it is left without burial.

One principal aim of the following book is to more fully articulate the nature of the unique importance of the complexity of human sympathy for animals by examining literature and art from the early modern Spanish world. A basic assumption in the following book is that, after greater exposure to the interdependency of the animal and the human, humans might, as Donna Haraway suggests, come to share the pain of the animal in order to promote thinking and action that minimize the exploitative ways that humans live at the expense of animals and that humans live with themselves.

By more fully introducing a dialogue between Animal Studies and the academic field of Hispanism,[5] the following book looks to not just Jacques Derrida and Donna Haraway, but also to Giorgio Agamben and Cary Wolfe.[6] As Wolfe has

---

to have a violent uncontrollable fit. Aside from *día de perros*, Spanish has an adverb that expresses that same negativity. To complain about how bad everything is going, one can say everything is going *perramente* (roughly, "like a dog" or "really badly"). In Mexico, the expression "¡Qué perro!" literally translates to "what a dog!", but roughly means "that sucks." *Perro amor*, a popular *telenovela* (soap opera) produced by Telemundo, uses *perro* as an adjective that modifies *amor* just as in the title of the film, *Amores perros*. In the case of the film and the television series, the adjective *perro* suggests "damned" or "messed up," making "messed-up love" a good translation of *amor perro*.

[5] Many scholars have questioned the name "Hispanism" as a name for the field. Some have proposed Iberian Studies (most notably Juan Ramón Resina) and others have proposed Trans-Atlantic Studies (most notably Joseba Gabilondo). I maintain the name of "Hispanism" because it best reflects the study of literature and cultures from the Spanish-speaking world.

[6] While I have found these thinkers the most useful in articulating the parameters of "Animal Studies," I do not mean to imply that they are *the* authorities of the Animal Studies approach. The list of scholars doing work in this field is too long to reproduce here. I should also mention that each of these Animal Studies scholars works out of a tradition that often has its roots in the medieval and early modern period. Derrida's interaction with his cat for instance (one of the cornerstones of "The Animal That Therefore I Am (More to Follow)" has precedence in *I and Thou* wherein Martin Buber looks into the eyes of the cat and construes the animal following the medieval tradition of placing it between the "vegetative" and "spiritual" (Kohanski 125). Derrida's cat comments also have precedence in Michel de Montaigne's assertion that that the superiority of human reason over animal instinct may be false: "By what comparison from them to us doth he conclude the brutishnesse he ascribeth unto them? When I am playing with my cat, who knows whether she have more sport in dallying with me than I have in gaming with her?" (159). While the early modern genealogical trajectory behind recent Animal Studies approaches is not the primary focus of this book, it is nonetheless significant and needs further study. Also see Fudge, "The Dog."

argued, Animal Studies has had an influence in radically altering the disciplinarity of numerous academic fields.[7] In the case of the field of early modern Spanish studies, this book studies its topic through the lens of the geographical scope of the early modern Atlantic world and Europe. Indeed, by including early modern England, it shows how anthropocentrism in Spain functioned in a similar way in a European context that both influenced Spain and was influenced by Spain.[8] Finally, in its interdisciplinarity, this study of the dog in early modern Spain does not limit itself to only the literary text or painting, but draws on a rich array of textual and visual sources in its analysis of each, including hunting manuals, sources related to animal husbandry, early modern dictionaries, and a film, among many other sources.

With respect to Animal Studies and early modern England, critics such as Laurie Shannon have looked to Shakespeare to point to moments of instability of the category of the human that so much of Renaissance sought to uphold and manufacture. Laurie Shannon has written how Shakespeare does not work within the period's ostensibly humanist bent, particularly with respect to popular philosophers like Giovanni Pico della Mirandola that conceive a world that centers on human capacity and perspective. Shannon writes with respect to *King Lear*: "Despite early modernity's reputation for an invention of humankind in terms of some new omnicompetence, *King Lear* exposes an abject humanity's underprovisioning in the face of the environment and its sheer incapacity for the great dramas of self-fashioning Pico had celebrated" (196).

Much of the underlying bent of Animal Studies on early modern England and the European Renaissance seeks to question critical efforts that have re-invigorated notions of the human and humanity.[9] In their approach to Shakespeare, critics such

---

7 Various critics understand Animal Studies as altering disciplinarity. One especially productive interdisciplinary approach to animals is that of the historian Susan J. Pearson and anthropologist Mary Weismantel who argue that for a new theoretical formulation with respect to animals "that incorporates symbolic approaches with social and material history" ("Does" 17). With respect to modern literary studies and critical theory, Carrie Rohman writes: "by attending to the question of the animal on a discursive and material level, we might begin to reshape our vision not only of modern literatures but also of the human and the humanities themselves. In doing so, we also answer our obligation to animals, who are looking at us now more than ever" (163). See Caesar for a study of the dog that employs Animal Studies perspectives similar to those in this book.

8 This book primarily concentrates on the parallels between the dog design in Europe and Spain—they both construct the dog in an othering-ennobling gesture. Nonetheless, it should be realized that Spain also radically departs from the rest of Europe in its treatment of the animal, particularly with respect to its belatedness with respect to taxonomy and other disciplines of zoology. Keith Thomas notes that one English clerk from the Royal Society reported that in 1701 the Spanish suspected that he was a mad-man because he had followed butterflies and gathered different types of flora (281).

9 See, for example, Schiesari, *Beasts*; Erica Fudge, *Brutal* "The Dog," "'The dog,'" *Perceiving,* "A Left-handed Blow;" Bruce Boehrer, *Cultural*, "Shylock," *Shakespeare*; and Laurie Shannon, "Poor," "Invisible."

as Shannon and Bruce Boehrer seek to find a critical vocabulary that does not simply return to the comfort of the nineteenth-century ontological subject that served the interests of nationalism and other human identity formations. Ever since the rise of post-structuralism in the late 1960s, the humanities itself experienced an identity crisis. In the face of that crisis, critics such as Harold Bloom returned to this ontological subject by clearly defining the object of study of the humanities (providing it a list and corpus of canonical texts) and by steadfastly arguing for a notion of the human. Harold Bloom writes in *Shakespeare the Invention of the Human* that Shakespeare creates the human insofar as he is the first universal author, that he replaces the Bible in secularized consciousness, that he informs the language we speak, and that his principal characters have become our mythology (718). As a response to Bloom, early modern English Animal Studies scholars perturb constructed notions of the human and humanity. Paralleling Shannon's purpose with respect to unsettling the notion of humanity in Shakespeare, Bruce Boehrer argues against the complacency and self-infatuation in critical efforts to celebrate Shakespeare "as the inventor of the human" ("Animal" 546).[10]

Even as critics such as Boehrer highlight early modern alternative paradigms taken from the English context, the largely Anglo-centric bias in Animal Studies needs revision since the Spanish proto-national and imperial moment compares to the English one, often serving as model and vice versa. Moreover, the following book introduces Hispanism to Animal Studies by unsettling the notion "humanity" with respect to two of its most celebrated figures, Cervantes and Velázquez. Paralleling Bloom's turn to the human in his study on Shakespeare, the figures of Miguel de Cervantes and Diego Velázquez have been especially important in seeking to re-invigorate a notion of humanity in twenty-first century criticism.[11]

[10] Marjorie Garber's position about Shakespeare and animals, specifically the dog, is slightly different. She argues that the critical return to the human is paradoxically taking place in the study of the dog: "At a time when modes of analysis like feminism, postructuralism, historical materialism, and identity politics all urge the situatedness of particular human subjects, calling attention to defining characteristics of difference, a popular nostalgia for humanism, for the universal 'us' of midcentury criticism, has led to a recentering of the canine in, and as, the canon. In short, Shakespeare, long in the popular estimation a guarantor of 'human nature' and or chief descriptor of 'the human condition,' is being used, these days, as the way of locating and describing true humanness—and humanism—in the place where it is paradoxically to be found: not among people but among dogs" (*Profiling* 184). Chapters 3 and 4 show that Cervantes and Velázquez not only turn to "popular" humanism, but to the sixteenth-century tradition of humanism and other "humanizing" traditions in order to forge a post-humanist vision. Moreover, in a curious parallel to Garber's conclusion about mortality in her chapter on "Shakespeare's Dogs," both Cervantes and Velázquez also turn to the dog when death approaches.

[11] See, for instance, Alban Forcione's introduction to his Lope de Vega study, *Majesty and Humanity: Kings and their Doubles in the Political Drama of the Spanish Golden Age*. Forcione argues that despite their biographical differences, both Cervantes and Velázquez have highly-significant points of commonality. With respect to the notion of "humanity," Forcione concludes that each recovers an unofficial reality, each invests it with a "thoroughly

Nonetheless, Velázquez and Cervantes represent the animal outside the design of human beings and human orderings and they conceive the animal, most especially the dog, outside of anthropocentric friendship.

Even as it often ignores the Spanish case, Animal Studies in criticism about Shakespeare and early modern England provides for a point of departure for this book's re-evaluation of the case of early modern Spain and how it unsettles humankind's sixteenth-century self-fashioning. The field of Hispanism has begun to feel the impact of Animal Studies.[12] Georgina Dopico Black argues that early modern Spain needs to be included in the Animal Studies debate because it,

> as a consequence of its imperial appetites and the political-theological hand-wringing that sometimes accompanied it—not only saw the rise of different border practices aimed at establishing (or, occasionally erasing) lines of demarcation between human and non-human (a constantly shifting category), but it was particularly interested in what lies at the edges of the human: the beast and the sovereign, but also the monster, the machine, the hermaphrodite, the native, the slave, and the divine. (237)

Borrowing the term from Donna Haraway, Dopico Black is especially interested in the "leaky" distinction between the human and its others (the animal in this case) in early modern Spain. For instance, Dopico Black analyzes a crucial ambiguity with respect to the animal in the work of the sixteenth-century scholar, Gómez Pereira. On one hand, Gómez Pereira ostensibly takes the humanist reading of Aristotle to show human exceptionalism by reading *De Anima* to prove the immortality of the human soul and also proposing animal automatism in anticipation of Descartes.[13] His work, on the other hand, is also a "rather rickety place on which to found a Cartesian subject" because his "mules wail and nod, his bulls bellow and

---

human" mystery and aesthetic, and each displays royalty in a process of divestiture of power that is simultaneously transferred to the "real human being" (2). Forcione's study of both Cervantes and Velázquez is an implicit response to Michel Foucault who also studied both Cervantes and Velázquez in *The Order of Things*. Foucault's scholarship on both Spanish figures helped to usher in post-structuralist approaches in the humanities in the 1960s.

12 In *The Animals of Spain: An Introduction to Imperial Perceptions and Human Interaction with Other Animals, 1492–1826*, Abel Alves studies how Spanish intellectuals regard animals, the Spanish empire's use of animals as labor and food, and popular beliefs about animals. Zeb Totorici and Martha Few have also edited a collection of essays on the history of animals in Latin America. The purpose of the study is to move away from an anthropocentric cultural history of animal by "centering animals." They write: "while much has been written on animals as symbols, representations, and phantasmagorical creatures, we situate the essays in this volume within a historiography of colonial and postcolonial Latin America that aims to make the interactions between human and nonhuman animals the primary focus of the analysis, so as ensure that the 'real animal' in question does not disappear from historical discourse" (n.p.). Thanks to Martha Few and Zeb Tortorici for letting me quote from their forthcoming work. For an introduction to Animal Studies and Cervantes, see Martin, "Zoopoética."

13 Also see Thomas 33.

bow, demanding mercy or protection" (247). Most significantly, Gómez Pereira mobilizes a sympathetic imagination toward the animal that "cannot be so neatly swept away" (247).[14]

When one begins to turn the humanist stones in the literature from early modern Spain, many more examples reveal leaky distinctions between the human and its other others. Aside from the New Mexico epic by Villagrá and the Gómez Pereira examples, one finds examples in Antonio de Torquemada's *The Spanish Mandevile of Miracles; Or The Garden of Curious Flowers* (*El jardín de las flores curiosas*); Luis Pérez's *Concerning the Dog and the Horse* (*Del can y el caballo*); and Antonio de Solís's poetry.[15] While mentioning these works, the following book is primarily centered on showing the unique way in which Miguel de Cervantes and Velázquez uniquely forge sympathetic imagination toward the animal.

---

[14] Compare Dopico Black's position about Spain with respect Keith Thomas's general position about the lack of human omnicompetence in general in the early modern world (his study focuses on England): "not everyone thought that the world was made exclusively for man, that nature was to be feared and subjugated, that the inferior species had no rights or that the differences between man and beast were unbridgeable" (50). Beyond Dopico Black's focus on early modern Spain, other general reasons can be cited for the need for Hispanism to engage Animal Studies, including: 1. The need to study the animal in the context of Hispanism in light of the fact that Animal Studies critics such as Cary Wolfe have cited Spain's interest in giving human rights to apes, but have not extensively explored the case. North American critics and contemporary writers in Spain such as Juan Mayorga are increasingly attempting to assess Spain's role in thinking about the animal as it pertains to larger debates across the globe (Beilin). 2. The need to think through the role of the Spanish language as "human" in the global economy. It is true that we may draw a parallel between Shakespeare as "human" with the figures of Cervantes and Velázquez, but more important is the need to assess in what ways the Spanish language and its cultures are constituted as "human" in the twenty-first century. As Shakespeare functions as human for Bloom or as the humanities functioned as the culture for science and nation states in the nineteenth century, so Hispanism often functions as a "human" culture of the global system. The United States is getting used to the fact that it does not just have one, but two national languages, Spanish and English (Alonso), and the field of Hispanism is rapidly growing in India and China. Many conceive Spanish as a humanizing language of the private sphere and a language of emotions and culture. The humanizing of Spanish is juxtaposed against English and the field of Hispanism is reduced to the "human" minority-language culture subservient to the global knowledge regimes that are largely English-speaking. The "minority" status of Spanish in this sense is thought to provide "humanity" to an a-human English majority (those global knowledge regimes). Animal Studies can provide productive theoretical terrain outside of the debilitating model that sets up the north as operating in English, the "language" of technology and science that "needs" the field of Spanish to supply a "humanizing" element.

[15] While I make mention of these texts in this study, more scholarship needs to be done on these texts and others, including those from the Americas such as the "Philosophical Discourse Concerning Animal Language" ("Discurso filosófico del lenguaje sobre los animales"), an unedited eighteenth-century work from New Spain that directly refutes Descartes and recognizes that animals can speak (Delmar).

Chapter 1, "The Hidden Dog," establishes the groundwork for an Animal Studies reading in the discipline of twenty-first century Hispanism by examining two influential scholars from last century. The animal is not an important object of study in the work of Américo Castro and Tzvetan Todorov, but it, nonetheless, must be considered in the evaluation of their work. In the case of Castro, *espagnol*, the word that played an important role in the origin of the modern word "Spaniard," is also the origin of a word for a canine breed, "spaniel." The conflation of the spaniel-Spaniard was already underway in the Middle Ages and became significant with respect to the formation of national identity categories in early modern Europe, especially England. In the case of Todorov, scholarship has been directly and indirectly influenced by his theory of the other, largely based on Spain's experience in the Americas. Todorov, however, does not consider Animal Studies in his work, an especially important lacuna since his theory of the other and humankind is marked by the symbol of the dog as the brutalizing force of the Spaniards. After identifying the hidden or forgotten dog in the work of Castro and Todorov, the chapter concludes with a nod to the dog in the work of Sebastián de Covarrubias, the important lexicographer from early modern Spain, in order to articulate how Animal Studies facilitates future paths for Hispanism.

Chapter 2, "A Cervantine Animal Exemplum: Animal Studies and 'The Dialogue of the Dogs,'" begins an analysis of the work of Miguel de Cervantes that continues in Chapter 3. Chapter 2 brings together themes from the dog dialogue with Animal Studies approaches, especially those from Cary Wolfe, Giorgio Agamben, and Donna Haraway. The argument turns to Animal Studies in order to argue on behalf of the elimination of the animal as figure in better conceiving the interdependency of animal and human. After showing how the dog dialogue is reflective of Cary Wolfe's critique of theories that assume that language belongs to humans, the chapter demonstrates how Cervantes ingeniously creates a new form of animal exemplum literature.

Chapter 3, "When the Dog is a Book: Reaching a Post-Human Ethics in Cervantes," articulates the notion of an Animal Studies ethics by studying the "Prologue to the Reader" of Part II of *Don Quijote*. The methodology of Chapters 3 and 4 is to examine how anthropocentrism operated in the early modern world and to show how Cervantes and Velázquez present alternative paradigms to anthropocentrism. In the prologue, Cervantes makes reference to and rejects three early modern anthropocentric representations of the dog: as part of a large-small discursive tradition; as object of carnivalesque ridicule; and as object toward which one should act humanely. Abandoning these representations and creating a dog-book (Cervantes literally equates his book with the body of a dog), Cervantes turns to an ethics that does not rely on the human-versus-animal since the dog-book demands an empathy that is constituted through the inclusion of non human faces that demand ethical responsibility.

Chapter 4, "As Death Approaches: The Dog in *Las meninas*," examines the figure of the Spanish mastiff that is in the process of being woken up in the complex narrative scheme of the painting. Velázquez's representation of the threshold figure of the dog in *Las meninas* is akin to Agamben and Wolfe's description of the animal as a living death. In the *Las meninas*, the figure of the dog connects to how King Philip IV conceives of his life and his own living death. In the representation of the dog, Velázquez departs from important sixteenth-century iconographic and anthropocentric traditions of the canine, including the humanist scholar with dog, the nobleman with dog, and the dwarf with dog. Instead, Velázquez associates the dog with approaching death. The dog in *Las meninas* is not a friend, that is, an appendage to the humanist that discovers truths that elevate humans toward the divine, nor a royal portrait companion that shows off the king's masculinity, faithfulness, and contained power. Instead, the dog reminds the king of certain truths that increasingly no longer hold, such as the central role of humanism in thinking toward the truth, the king's masculinity and future progeny, and his empire's military prowess and wealth.

Broadly, this book is about canine design. Over two hundred years before England standardized breeds in the nineteenth century, England (and Europe in general) had developed various types of spaniels. Moreover, sixteenth- and seventeenth-century Spain designed over twenty different breeds that connoted a notion of dog pedigree versus the mongrel.[16] This study thereby provides the first comprehensive history of early Spanish dogs. It focuses most especially on three of these Spanish dog designs: in Chapter 2, the *alano*, the large game hunter and bull-baiter; in Chapter 3, the *podenco*, a medium-sized rabbit hunter; and in Chapter 4, the *mastín,* the unique and immense Spanish sheep dog. The book is also about literary, artistic and aesthetic "dog design" since it examines how and why Cervantes and Velázquez created dogs.

---

[16] For instance, a litany of dog breeds are used as comic relief in the *entremés* "La rabia" by Pedro Calderón de la Barca. In "La rabia," a pun on "rabies" and "anger," one character is bitten by a *braco* or pug variety. She exclaims: "I had just knocked when the door was opened / and a whole army of dogs came out / a little mutt, a water dog, a pointer, / a poodle, a hairless, and a lapdog; / a Spanish mastiff, a rabbit dog, a greyhound, / a Great Dane, a bloodhound, a scent dog ... (Oh! I can't breathe!) / and a nasty little pug ... ("Pues apenas llamé, cuando al abrilla, / a la puerta salieron en cuadrilla / un gozque, un perro de agua, un perdiguero, / un lanudillo, un chino y un faldero; / un mastín, un lebrel, un galgo, / un dogo, un sabueso, un ventor ... [¡Ay que me ahogo!] / y entre ellos un ladrón de un perro braco" [my trans.; Calderón]). In another play, Lope de Vega describes one suitor who gave different types of dogs to different types of women. He gave "lapdogs to the humble / pugs to the prudish / rabbit dogs to the proud / and Spanish bulldogs to the crafty" ("a las humildes, de falda: / a las melindrosas, bracos: / a las soberbias, lebreles, / y a las taimadas alanos" [my trans.; qtd. in Fichter 144]). Many early modern hunting manuals list dog breeds and provide descriptions. Alonso Martínez de Espinar's manual (1644), for example, describes the rabbit dog, the bulldog, the Spanish mastiff, the greyhound, a French and Spanish blood hound, a pointer, a water dog, a *podenco*, a vermin dog, another type of pointer, and a fox dog (*lebrel, alano, mastín, galgo, sabueso de Francia y de España, de muestra, de agua, podenco, de alimañas, de encarbo, zorreros* [my trans.; 61–2]).

Following the Afterword which discusses the film *Amores perros*, the Appendix offers an introduction to the topic of breed and race with the hopes that it will stimulate future research. "Design" does not only mean the way in which people (hunters, sheep herders and early modern pet owners) favored certain physical features on specific canine varieties. A basic premise of the following is that, as dogs are shaped, so their shaping inevitably implicates a reciprocal shaping of the human. This reciprocal relationship is not only one in which the hunter shapes his dog as the dog shapes the hunter, but also, as breeds developed, so did conceptual notions of the human such as the development of early notions of race. The Appendix, building on the same argument from Chapter 1, historicizes the othering and ennobling sign of the animal. By knowing the etymologies of human identity terms that originally meant animals in the sixteenth-century Spanish Atlantic world, scholars will better recognize how anthropocentric assumptions inform definitions of humanity, more often than not based on an unproductive outside-versus-inside model.

With the idea of a reciprocal relationship—human design of dogs and dog design of humans—this book is concerned with the history of affect. Particularly, the exploration of the animal as "other" other exhorts scholars to be mindful of affect. How does one "suffer with" (that is, have "compassion" or "sympathy" towards) the dog or the animal in general? In "The Friend of My Friends," Dominique Lestel considers the unique role of the animal as a different other:

> We have not paid sufficient attention to this recurring fact, that it is extremely difficult for us, we other humans, to remain indifferent to the animal. The question of suffering, such as utilitarians like Jeremy Bentham pose it from the end of the 18th century, is coupled with that of love and friendship, something that is neglected by the philosophers open to the transhuman—starting with Bentham himself. Not only: can animals suffer? But also: can animals love? And, more precisely: can animals love us, us humans, as we love them? A certain *cognitive anthropomorphism* loses pertinence to *affective anthropomorphism*. n.p.[17]

One reason that the study of the dog is an especially good point of departure for the study of the affect is because it—from Odysseus' Argus up until Lassie—is the animal *par excellance* that supposedly "loves" its human companion. Nonetheless, its role as human "other"—though apparently obvious—is not at all sufficiently understood. Purportedly man's best friend, most would agree that the dog can suffer. The dog, perhaps more than any other animal, has given the human who chooses to interact with it a fulfilled, blissful feeling. But it also might always *turn* on its owner and steal that piece of meat you were hoping to have for supper. It does not have human language or a human face. It is not the "other" or subaltern

[17] Thanks to Jeffrey Bussolini for giving me permission to use his unpublished translation.

as theorized in postcolonial studies.[18] Like the dog's body in *The History of New Mexico*, the nature of its memorial, the way in which its death is marked and the way we remember the dog, is still very much an open question that demands consideration. The following sets out on the task of tackling this open question by examining the dog as protagonist in early modern Spanish culture.

[18] Some scholars assert that the animal is an "other" or "subaltern" following theories from post-colonial studies. One reason that I argue the animal is an "other" other is to distinguish the way in which Animal Studies uniquely defines the animal as distinct from the other or "subaltern." In other words, this book does not treat the animal as an "other" that recuperates its voice and narrates its "own" life. Animal Studies provides a theoretical discourse outside the logic of the ontology of race-thinking and anti-race thinking alike because only outside of the anthropocentric base of the semantics of a more-or-less animal argument will the human not be the end of subaltern thinking. The human in Animal Studies is not based on the supposition that it is constituted through its negation with the animal. Animal Studies reminds the critic that if thought is also not directed toward the animal, then thought fails toward those humans who are othered under the sign of the animal. Graham Huggan and Helen Tiffin's work on ecocriticism, literature and the environment argues that human liberation will never be fully achieved without challenging how human societies have constructed themselves in hierarchical relation to other human communities through the non human. Huggan and Tiffin warn against the animal-human hierarchical entrenchment perpetuated by arguments for genocide as well as those against. Societies employ the animal to marginalize other human societies and human genocide is predicated on categorizing people as animals. In turn, the condemnation of genocide is based on an animal-human divide, fixing a hierarchical boundary between the human and animal since humans are *not* animals and should not be treated that way. In the condemnation of that marginalization and genocide, societies ironically return to the animal-human divide argument by stating that because humans are *not* animals becausethey should not be treated that way.

Chapter 1
The Hidden Dog

> *Canes*, when part of buildings, are either wooden or stone heads upon which long beams are placed to mount the floor…And they call all of these *canes* because they are typically carved into the shape of the heads of dogs. These dogs are also called *atlantes* because they are like the figure of Atlas who is painted holding up the heavens; and in this way the *canes* provide the structural support for the building. (my trans.)
>
> Canes, en los edificios, son unas cabezas, o de maderos o de piedra, que suelen sobre ellos sentar unas vigas largas, y armar los suelos sobre ellas…Y llamáronse todos éstos canes porque ordinariamente están labrados en ellos unas cabezas de perros. A estos canes llaman atlantes por la semejanza que tienen con Atlas cuando le pintan sustentar el cielo; y así los canes parecen sustentar el edificio. (281)
>
> —Sebastián de Covarrubias

Cary Wolfe has argued that scholars must question the human that undertakes pluralization with respect to animal and human others in order to end the anthropocentric base of the "more-or-less animal" argument in which "humanity" is the end of thinking about the other. In formulating this argument, Wolfe follows Jacques Derrida who proposes the need to think a different other, an "other" other, that is not based on an animal–human differential. This "other" other is, at its root, non-oppositional and infinitely differentiated—not reduced to the "inside," that is the general human subject versus the "outside," that is, the non-subject that is animal. A fuller articulation of this new other ("significant" or "other" other) that follows the Wolfe, Haraway, and Derrida Animal Studies model will help to foster critical approaches that focus on a non-oppositional other no longer bound to identity formation based on region, race, nation and gender.

I will carry out the work of further delineating this Animal Studies other in the following three chapters in this book. In turn, in this first chapter, I set the stage for the last three chapters by elucidating the inside-outside anthropocentric model of the other in the work of Américo Castro and Tzvetan Todorov, two thinkers who had an important impact on twentieth-century Hispanism. An analysis of the omitted or "hidden" dog in the work of Castro and Todorov aids in revealing how the work of each participates in an inside-outside anthropocentric model. This chapter sets the terrain for the last three chapters of the book and also for future scholarship on Animal Studies in Hispanism in the sense that critics, after the historicization of the othering and ennobling signs of the animal, will more easily recognize and discard unquestioned anthropocentric assumptions in their scholarship. That historicization will aid scholars of the early modern period to cast off a definition of humanity that depends on identity formation based on an anthropocentric model of the outside versus the inside.[1]

Américo Castro was one of the most influential scholars in the twentieth-century because he emptied out the essentialist character of the Spanish national identity category (Cruz 84). As Castro writes, the word "Spaniard" (*español*) was not a term from time immemorial, but acquired significant currency for the first time in the sixteenth century when it appeared in the title of books beginning in 1520. At the time of the conquest, the conquerors did not principally conceive of themselves as Spaniards, but first as vassals of the king, as Christians, and members of smaller regions. (The primary language at the time of the conquest was not conceived as *español*, but as *castellano* and for that reason *castellano* not *español* is still preferred as the term used in many parts of America.) (43). Américo Castro explains that "Spaniard" did not originate in the Spanish lexicon from the inside (the Iberian Peninsula) since the Spanish had no national, political, or linguistic reason for considering themselves Spaniards (*españoles*). Instead, an outside region imposed the word on the Peninsula—the extra-Peninsular neighbor, Provençe, used the word "Spaniard" (*espagnol*) to describe its neighbor to the south.

An Animal Studies approach considers the Castro method—the historicization of the human category of "Spaniard"—but it also takes into account the history of the animal and the way in which early modern Europe employs an anthropocentrizing gesture when it conceives Spain from the outside. In *Spain Beyond Spain: Modernity, Literary History, and National Identity,* Brad Epps and Luis Fernández Cifuentes edit a collection of essays that propose "another take" on Spain "to represent a diversity of positions" (20), including cultural studies, postcolonial studies, new historicism, women's studies, ideology critique, and trans-Atlantic studies. As an epigraph for the collection of essays, Epps and Cifuentes choose a quote from Américo Castro that reflects how Castro struggled

---

[1] This chapter ultimately envisions an Animal Studies theoretical discourse that discards the inside and outside. Its goal parallels those expressed in studies by Margaret Greer and Alberto Moreiras. Margaret Greer looks to future scholarship when studies of early modern Spanish period are neither on the "inside" nor "outside" ("Mine"). For Alberto Moreiras, there is no longer an imaginable outside and no possibility of an inside with theoretical practice that imagines the friend as the absolute subsumption of otherness. Moreiras—in a parallel gesture to Wolfe and Derrida's Animal Studies perspective that seeks to look beyond the human—argues that theoretical practice itself is the new other: "critical thought or an unprecedented thought is when we come to call the friend by a name that is no longer that of the near one or the neighbor, and undoubtedly no longer the name of man" (*Exhaustion* 293; see also Derrida, *Politics* and Žižek). In an article on the future of Hispanism, Moreiras draws on his theorization of "friendship" (a particularly useful motif for studies on the dog). In the collection of essays entitled *Ideologies of Hispanism*, Moreiras turns to the animal, specifically to a mule ("Mules"). In discussing the human walk with a mule from a Lezama Lima novel, Moreiras comments on how the human and the mule walk side by side, but the mule no longer is the beast that gives the human its name.

with the meaning of Hispanism—"Spain," the "Spaniard," and things Hispanic—as a cultural colony of the foreigner because they have been an interpellation by those from the outside.

The miscellany of approaches in the Epps and Cifuentes volume are intended to offer twentieth-first-century Hispanism fecund approaches to colonization, globalization and identity, breaking from the history of the field that once traumatically recognized itself as a construction by the outsider. But in their consideration of the future of Hispanism, Cifuentes and Epps do not consider Animal Studies, an especially needed approach given the Castro quote in the epigraph about the "outside." Castro rightly pointed out the genesis of the identity category "Spaniard" (*español*) as largely an outside imposition, but he did not consider how much the animal, specifically, the Spanish dog played a role in the interpellation of *español*. "Spaniard" (*espagnol* in Provençe) was not just a designation from the outside (those from Provençe) on a human group on the inside (the Spaniards), but *espagnol* also meant "Spanish dog" in Provençe. Therefore, the Provençe word *espagnol* meaning dog *and* human moved south to become the word *español*. The region outside of Spain ennobled itself through the othering of Spain by making the *espagnol* not just the human Iberian inhabitant, but also the sign of the animal.[2]

The nationality assigned to dog breeds often reveals significant information about cultural history. The creation of the German shepherd in the 1920s is an example of a canine dog design connected with nationalism and racial identity since it was crafted at the service of Nazi ideologies (Skabelund, "Breeding").[3] The case of the Chihuahua is symptomatic of how the foreigner is absorbed into the domestic space of self in the formation of national identity. When H. Raynor of El Paso, Texas recorded the first Chihuahua with the American Kennel Committee on July 18, 1903, the animal could have been called a "California," "Texas," or "Arizona" dog (names used by many in the late nineteenth century for the tiny

---

[2] Etymologists also included the animal into the spaniel's Spanish origin by connecting the etymology of "Spain" with rabbits. When Virginia Woolf wrote the biography of Elizabeth Barrett Browning's cocker spaniel Flush, she repeated the commonly-held etymology that *Span*, the word that gave the word Spain, connected to rabbits. For her, *Span* was an old word for a rabbit or a rabbit dog. Woolf explains: "The [Iberian Peninsula] was alive with rabbits. And *Span* in the Carthaginian tongue signifies Rabbit. Thus the land was called Hispania, or Rabbit-land, and the dogs, which were almost instantly perceived in full pursuit of the rabbits were called Spaniels or rabbit dogs" (11–12). Woolf confirms an anthropocentric assumption that surfaced in early modern Europe in which the name "Spain" was etymologically connected with the animal: "we," Europe in this case, the humans on the inside name those "animals" on the outside.

[3] Toni Morrison provides a parallel example to the creation of the German shepherd when she writes of how the white man's "friend" is an actor in racism. She writes in the *Song of Solomon* a description of white sympathy for dogs, but hate toward blacks: "White people did love their dogs. Kill a nigger and comb their hair at the same time. But I've seen grown white men cry about their dogs" (51–2).

canine), but the name of the bordering Mexican state of Chihuahua was chosen (Fernandez 4).[4] The state of Chihuahua—versus California, Texas and Arizona—of course is a state in Mexico, not in the U.S. Despite its connection with the state of a foreign nation, the Chihuahua dog was introduced inside the intimacy of the space of the US home. Attesting to the popularity of the dog in the US household, ownership of the Chihuahua became so popular that it resulted in a canine glut in California animal shelters, resulting in the "Great Chihuahua Airlift of 2010" in which private individuals and organizations flew the surplus Chihuahuas to New York (Metzger). Closely associated with what many Americans call their "own" and part of the affective space of the home, the Chihuahua is also sometimes an icon for food and fashion such as in the case of Taco Bell's Chihuahua and Paris Hilton posing with her Chihuahuas.

Most canine enthusiasts consider the development of the spaniel design to be largely a nineteenth-century British invention in which the dog is especially associated with England.[5] Likewise, the connection between breeds and nationality is generally considered to arise in nineteenth-century England such as the appearance of a series of cartoon caricatures in which one finds the no-nonsense bulldog, associated with iconic English virtues of courage and tenacity, contrasting the French as small prancing poodles (Johns 136).[6] But the nationality assigned to the early modern European dog breed, the spaniel, in the sixteenth century already reveals how dogs connected with nationality, a significant ellipsis in Castro's analysis of the origin of *español.* A study of the spaniel more fully elucidates Europe's perception about Spain, the nature of the formation of proto-nationalism, and about the construction of gender in early modern Europe. Just as those who own a Chihuahua do not consciously animalize or make a pet out of

---

4 As one internet source writes: "Some believe that it originated in Asia, others that it came from Europe. But no matter where the breed got its start, it really is a dog of Mexico" ("Chihuahua"). Dog enthusiasts often claim the origins of the Chihuahua to be pre-Columbian breeds like the Mexican hairless, often referred to as xolos. Since, however, no historical record of dwarf dogs exists in Mexico until the late nineteenth century, most likely, what is billed as "the world's smallest dog," was an Asian introduction into Mexico.

5 Modern dog breeds first date with the establishment of the English Kennel Club in 1873 and the subsequent documentation of pedigrees. For a general history of the development of breeds, see Hancock. The spaniel as lexical item and as physical dog became the property of the nineteenth-century British in the sense that spaniel varieties were designed as British. Although there would eventually be a few non-British spaniels (the American Water, the Irish Water, the Tibetan), the spaniel and its varieties (the Cocker, English Cocker, Springer, Clumber, Field, and Sussex) became England's favorite and *most* beloved. These spaniel varieties created in the nineteenth century were based on early modern hunting spaniels. The nineteenth-century toy varieties (including the English toy and the Cavalier King Charles) were based on the sixteenth-century "gentle" or "noble" spaniel types.

6 For recent examples, such as the connection between the bulldog as symbolic of British political power, see Baker 35.

the northern Mexican state that borders Texas and New Mexico, so the sixteenth-century Europeans, especially the English, who owned spaniels, did not literally see themselves as converting the Spanish into little lapdogs. Nonetheless, similar to the coining of the Chihuahua's name as a North American dog, so the evolution of the spaniel into an English dog forms part of a cultural turn in which things Hispanic and Spanish constitute a pet-keeping practice that symbolically provide the pet owner cultural caché. Just as many in North America love the Chihuahua dog, having named the outsider its own, so England loved its small spaniels. Both the Chihuahua and the spaniel's small size and invented names connect with the gendered and national deployments of the dog, attesting to the appropriation of the canine in shaping national identity and gender roles.

Paralleling the case of the Spanish dog, the nationality assigned to horse breeds reveals socio-cultural fears and desires in early modern England. Just like the Spanish dog, the Spanish horse in England was portrayed as the cleverest of its kind and the English home became the "natural" haven for the Spanish breed. England had domesticated the Spanish horse as its own, taking the sign of the purebred quality of the Spanish horse and turned it into an ennobling sign. By owning a Spanish horse breed, the early modern English performed an act of domestication of a feared imperial power that materially confined the animal, the horse, and symbolically confined the Spanish as part of the English landscape and home.[7] Through the material and symbolic sign, the English could at once indirectly praise the Spanish (the Spanish horse was considered the most intelligent of all horses), but also assimilate the Spanish as a part of an English national character since the Spanish horse was controlled and owned by the English.[8]

Castro's spaniel oversight is crucial because it forgets that early modern Europe also included the sign of the Spanish dog with the human in a discourse of othering and ennobling process constitutive of the human. A region outside of Spain ennobled itself through the othering of Spain by making the *espagnol* both the human Iberian inhabitant and the sign of the animal. As an interesting precedent to Cervantes' dog dialogue, in 1537 Bonaventure des Périers wrote a dialogue between two dogs. One of the dogs discusses different breeds and he mentions that he has a friend that is a spaniel. He uses the word *espagnol* (79). Thirteenth-century French iconography and literature had favored a dog called the *brachet* that sat at the back of its master's horse, pointed to the game, and retrieved it. But in the fourteenth century the spaniel (*espagnol*) replaced the *brachet* as

---

7 It should be noted that early modern European horse breeds are often associated with outsider regions such as the Barbary breed.

8 Karen Raber studies the Spanish horse to show how the seventeenth-century English home offers a haven for securing the Spanish. English horsemanship treatises portrayed the Spanish horse as unnervingly intelligent. In *A New Method and Extraordinary Invention to Dress Horses* (1667), William Cavendish writes that "You must know that of all horses in the world, of what nation soever they be, the Spanish horses are the wisest ... and strangely wise, beyond man's imagination" (qtd. in Raber 236).

the favored hunting breed (Bugnion).[9] The term *espagnol* could be defined in two ways: either as "Spaniard" or "Spanish dog" in which the word "dog" is understood, but elided. In *The Book of the Hunt* (*Livre de chasse*, 1387), Count Gaston Phébus, writing in Provençal, described the *espagnol* as a dog that worked by quartering in front of the hunter, and then flushed rabbits for the hound or game for the falcon. He also described how the *espagnol* retrieved ducks and geese from the water. He writes that this type of dog has the name *espagnol* because it comes from Spain (Miramon 208).

Significantly, the coining of the name of the Spanish dog occurs just as the late medieval Europeans increasingly adopted the sign of the dog as an ennobling sign. In the earliest medieval hunting manuals, hunting birds are ennobled (Albertus Magnus lists ten types of noble falcons and ten types of common ones), but the ennoblement of different types of dogs increasingly took place in the fifteenth century most especially with the emergence of the spaniel breed (Miramon). This ennoblement tied the sign of the animal to national identity and othering. In the case of the *espagnol* ("Spaniard" *and* "Spanish dog") the hunter ennobled himself by owning the Spanish dog, but also uses the spaniel to lambast the Spanish character. Indeed, even as he uses them as hunting dogs, Phébus also disparages the canine version of the *espagnol* in his *The Book of the Hunt* because it is no different from the human Spaniard. He links national character to race when he writes that the Spaniard, both the dog and human, are brawlers and howlers and their bad nature derives from their bad "generation" (Miramon 208).

"Cocker spaniel" entered the Spanish language as a loan word from English in the twentieth century (Wessem), but the spaniel was never a Spanish breed in the sixteenth and seventeenth century. The word *espagnol*, however, was common

---

[9] In Spanish, *brachet* would translate to *braco* and this breed today refers to various pointer or setter varieties. Semantic variations on *braco* (such as *brachet*) can be found in medieval English and French literature, where the word generally means a small hunting hound and pre-dates the emergence of the spaniel. In Spain in the seventeenth century, the small dog known as the *braco* had all the characteristics of the modern day pug, often described as having a cut-off, upturned muzzle. Authors like Francisco de Quevedo made jokes that connected the *chato* or *romo* with the *braco*. A *chato* or *romo* (see Chapter 2) meant "a snub-nosed person" and the words implied a person whose face had been damaged by syphilis or they were used as epithets of blacks in general. In one play, *His Sister's Brother* (*El hermano de su hermana*) by Francisco Bernardo de Quirós, one character calls Zoraida the Moor a "perra braca." But even though authors used the *braco* to disparage Moors, women, or blacks, it was thought to be a breed that came from the north. In a similar nationalizing semantic gesture as the *galgo*, the origins of the dog are set in France. The etymologist Sebastián de Covarrubias, after stating that the *braco* is a small breed with large ears that droop over its face, explains that: "This breed came to Spain from France and the word is French" ("Esta casta vino a España de Francia" y "es nombre francés" [my trans.; 233]). In contrast, English etymologists did not look to France for the brachet's origin, but to Scotland (Topsell). Just as England makes the greyhound's origin from Ireland, so it also appropriates a contested territory into its national space through assigning the brachet's origin to Scotland.

in medieval dialects of French, and in early English. Just as the Provençal word *espagnol* moved south into medieval Spanish and became the word *español*, so it also had moved to the north and became the English word spaniel (the epenthetic "e" from French and Spanish is simply dropped in English). Just like the doubly semantic French word *espagnol*, "spaniel" in early English did not just refer to a dog, but also meant "Spanish" (as an adjective) and "Spaniard" (as a noun) (Oxford English Dictionary). Writers as early as Chaucer, drawing from classical tradition that debased the small dog, wrote about "spaniels." The licentious Wife of Bath "coveteth every man that she may see; / for as a spaniel she will on him leap" (Chaucer).

In an act of anthropocentric ennobling, the English owned the Spanish by owning the dog, building a specifically English national and masculine character. In the early modern period, the spaniel would be confirmed as a distinctly British animal, whose "deep" origins were located in Spain. The English dog scholar John Caius (1570) comments that the origin of the meaning of spaniel takes root in the fact that it came from Spain: "these kinde of Dogges came originally and first of all out of Spaine" (15). Compiling a massive history of animals, Edward Topsell (1607) helped to create the attitude found for the centuries to come that the spaniel may indeed come from Spain, but that it properly belongs to England. Topsell writes the spaniel is "naturally" bred in England: "Spaine, Hispania, the country from whence (the spaniels) came, not that England wanteth such kind of dogs (for they are naturally bred and ingendred in this country), but because they bear the general and common name of these dogs since the time they were first brought over out of Spaine" (180).

Increasingly in the sixteenth century, spaniels in Europe, especially England, were not only a hunting variety, but also served as companion animals. The sign of the spaniel ennobled the English through the creation of a distinctly English masculinity by way of a nationalizing the foil of a feminized spaniel-Spanish character. In contrast to the relative paucity of medieval references, descriptions of the spaniel pervade sixteenth-century English texts, indicating that English pet-keeping was tied to the female gendering and eroticization of the dog and the Spaniard.[10] A common spaniel proverb from the period, used to justify abusiveness

[10] Shakespeare provides the best sampling on the usages of the word "spaniel" in the period. The Bishop of Winchester tells the King in Shakespeare's *Henry VIII*: "But know I come not / To hear such flattery now... / To me / you cannot reach. You play the spaniel, / And think with wagging of your tongue to win me..." (Act V, Scene 3). The ironic Petruchio in *The Taming of the Shrew* names his spaniel Troilus because of its legendary uncritical attachment to love objects. In *Julius Caesar*, Caesar criticizes "base spaniel-fawning" (Act III, Scene 1). In *Two Gentlemen of Verona*: "Yet, spaniel-like, the more she spurns my love, / the more it grows, and fawneth on her still" (Act 4, Scene 2). In *A Midsummer Night's Dream*, Helena tells Demetrius that she is a spaniel: "And even for that do I love you the more: / I am your spaniel; and Demetrius, / the more you beat me, I will fawn on you. / Use me but as your spaniel—spurn me, strike me, / Neglect me, lost me; only give me leave / (Unworthy as I am) to follow you. / What worser place can I beg in your love / (And yet a place of high respect with me) / Then to be used as you use your dog?" (Act II, Scene 1,

towards women, also suggests an English fantasy about the Spanish: "The more a spaniel is beaten, the fonder he is." [11] Following up on the Wife of Bath literary precedent, images of spaniels and the spaniel's tongue were associated with sex. Marjorie Garber and Ian MacInnes list early English sources in which spaniels are figures of male fantasies in that they represent the insatiable sexuality and sexual duplicity of women. A host of male poets connect spaniels with sexual envy, wishing they, humans, could perform the services that a dog performs. One poem is suggestive of the spaniel's privileged place in its mistress's lap: "The little spaniel in the Lady's lap / is blessed with extraordinary hap, / feeding and lodging in that princely place" (qtd. in MacInnes 36–7). One source is more direct, recording clients at a brothel who would love to "just to kiss the part where pleasure lies, like spaniels lick the centre of all joys" (qtd. in Garber, *Dog* 143).[12]

The semantics of the spaniel indicate an unconscious English masculine sexual desire and a desire to tame the dog, domesticating the rival imperial power, the killer dog, into a small fawning creature.[13] Following the precedent of Gaston Phébus who linked both the Spanish dog and the Spanish people as loud and unruly, sixteenth- and seventeenth-century English descriptions of the spaniel resonated with English descriptions of the inhabitants of Spain as servile, lust-ridden, inferior, and animal. Generally these references are only indirect inferences, but some English writers directly connect the spaniel to the Spanish people. In one case the spaniel–Spaniard connection stems from the English–Spanish religious rivalry. One writer calls Catholic Spaniards "spaniels" when he writes of the "Pope's spaniels" (qtd. in MacInnes 35). Another connects the behavior of the dog with the behavior of the Spaniards and another animal (the fawn) when he writes of the "fawning spanolizing Spaniel" (qtd. in MacInnes 35).[14] Paralleling the way in which they design the canine spaniel both physically and rhetorically,

---

ll. 199–210). Also see Jackson. In a feminist reading of the dog, Melissa E. Sanchez reads the trope of the spaniel as associated with a type of female masochism "that is not always a pathology that enables patriarchy but can be a legitimate form of desire that challenges traditional ideas of normal and proper female behavior" (493).

[11] For more on the common English proverbial phrase, "to fawn like the spaniel," see Jackson.

[12] For a depiction of a small dog whose tail explicitly covers and substitutes female genitalia, see the sexually-charged *Girl Making a Dog Dance on Her Bed* (late 1760s) by Jean-Honoré Fragonard.

[13] Compare the English design of the spaniel and the "dogging" or disparagement of Spain with Barbara Fuchs' study on how the English steal away Spanish cultural capital by discrediting Spain and its literature, especially Cervantes. Fuchs has argued that England occludes Spain and she points to the English national investment of literary history that demolishes Spanish glory. When England created the spaniel, they usurped the Spanish imperial canine and debased the Spaniard.

[14] The spaniel connoted feminine subservience, false friendship, sexual supplication, and cringing. This meaning was often achieved by equating the spaniel with the fawn, the interchanging of one animal simile for another.

the English, as the sixteenth century advanced, especially after the defeat of the Armada, increasingly conceived the Papist Spaniard as a fluffy ornate creature who ran around the royal court, but held no real power.

The first moments when the Spanish began to consider themselves Spanish—the initiation of the ennobling of the "Spaniard" in the sixteenth century—formed part of an anthropocentric cycle in which Europe disparaged Spain an animal-national other. Aside from the discourse of national identity and gender, Europe's othering of the Spaniard through the sign of the animal was tied to early notions of race. A series of early modern texts associated human physical features with those of animals, such as Bartolomeo della Rocca's (also called Cocles) Latin treatise on the subject in 1503. Bartolomeo della Rocca's text, published and translated in Europe through the sixteenth century, described human characteristics in relation to body parts and relates those body parts to certain animals, writing that the likeness of human and animal features results in similar nature.[15] The "Spaniard" and "Spanish dog" (*hispaniolus*) had a role in Rocca's physiognomy theory. Anticipating the nineteenth-century science of phrenology by centuries, Rocca quite literally implants the "Spanish" features of the dog on to the human whose head supposedly resembled the Spanish dog's head. Rocca borrows from the early modern assertion that small dogs bark a lot and connects it to a person with the shape of a spaniel's head, suggesting that such a person talks a lot and is quick-tempered.[16]

---

15 Bartolomeo della Rocca's text (translated to English by Thomas Hill in 1556) associates physical parts of the body, including one's hair, shape of the head, eyebrows, eyelids, eyes, nose, lips, tongue, mouth, voice, breath, laughter, color of the body, shoulder points, arms, breast, hands, and "the belly, back, greate guts, and haunches, with the legs to the feet" with essential aspects of a person. He writes that "the chin round declares that person to be effeminate, conditioned like the woman..." (n.p.). Rocca describes human characteristics in relation to body parts and relates those body parts to certain animals, writing that the likeness of human and animal features results in similar nature. Ears signify the condition of an ape or teeth like a dog's mean dog qualities: "They which have oversmall ears, be conditioned to the ape....The teeth longe like dogs' teeth, and strong set and properly bearing outward, declare that man to be a very gluttonous person and fierce, angry, compared to the dog" (n.p.). Other texts also associated human physical features with those of animals, such as Jerónimo de Cortés' sixteenth-century treatise *Physiognomy and Various Secrets from Nature* (*Fisonomía y varios secretos de naturaleza*) and Giambattista Della Porta's *Concerning Human Physiognomy* (*Della fisononmia dell'uomo*) (1586). Della Porta linked a narrow human face to a swine's snout, indicating a slow-witted and filthy man (Hill, *Casta* 248). Also see Schierasi 54–72.

16 The Rocca text reads as follows: "Plato, the wisest, writes that likeness of human and animal features results in similar nature. Who has an eagle-nose is noble, cruel, and predator likes an eagle. People having their heads like a Spanish dog are quick-tempered and talkative" (qtd. in Gale).

Early modern Europe designed an infantilized Spanish dog.[17] Since the early sixteenth century, Italian, Flemish, French and English painters depicted Spanish dogs or spaniels, the breed that gained increasing popularity across Europe since the fifteenth century. Those spaniels ennobled their European masters, representing wealth, power and exclusivity. Breeding practices associated with the spaniel also represented Europe's fantasy of how it conceived and wished to conceive the Spaniard. In sixteenth-century European courts, the spaniel was bred smaller and smaller, just as the word "spaniel" was used more frequently. The smaller the size, the more valued the spaniel. One early English source describes the spaniel as follows: "These puppies the smaller they be, the more pleasure they provoke, as more meete play-fellows for minsing mistresses to bear in their bosoms, to keep company with all in their chambers, to succour with sleep in bed." (Topsell 172; also qtd. in Thomas 107–108).

Portraiture examples of the favorite features in the breeding of the spaniel reveal a canine design symptomatic of the desire to see it shaped into a child-like plaything. The diminishing dog developed finer bones and profuse ear feathering. Titian's spaniels had relatively flat heads with a little stop; a type of toy spaniel painted shortly after by Veronese and others had high-domed, sometimes bulging heads. Generally, the small Spanish dogs had drooping ears. The ears were set high, although far enough apart to show the curve of the skull (Kemmerer). Europe designed the animal, especially its head, in correlation with its perceptions of the Spanish people, reflecting a connection between physiognomy and national character. The breed would also become known as a "dwarf spaniel." In early portraiture the connection between the small dog and the Spaniard was not direct, but not entirely absent either. Sixteenth-century painters depicted the spaniel with small people, such as children in Titian's *Clarice Strozzi* or dwarfs in Edward Tiberio de Tito's *A Dwarf with Medici Dogs in Boboli Gardens* (1620–25).

The scholar Juliana Schiesari looks to the "reclusive corners" of the domestic world of Italian renaissance—both the invention of the domestic space and the development of humanist thought. The case of "Spanish dog" as spaniel reveals one such reclusive corner found in European, but *not* in Spanish, households. Early modern Spain reveals important features of homes built on animal breed names and proto-nationalist formations. Greatly expanding the late medieval use of the dog as

[17] The breeding of dogs away from wolves favored infantilizing features: broad forehead, floppy ears, and shorter snout. Scientists attribute this to an inherent need to nurture in humans. Salisbury writes: "the main characteristic of all domestic animals is pedomorphosis, that is, the retention of juvenile characteristics, both in body shape and in personality characteristics, such as whining and submissiveness. The most extreme example of the retention of juvenile traits comes in toy dogs: in addition to their small size, they have disproportionally broad heads, small limbs, large eyes, and smaller noses and mouths. All these are characteristics of human infants and thus evoke what Nobel Prize-winning ethologist Konrad Lorenz defines as the 'cute response'" (116). Assuming the "cute response" is universal, scientists have not considered how it is indebted as a breeding technique to the constructions of race, gender and nationalism in the early modern period.

an othering and ennobling sign, sixteenth century Europe—especially England—made the Spanish "spaniels" and its spaniels "Spanish." The spaniel gained the reputation of being England's first and most endearing pet. Spaniels were always acceptable gifts between English aristocrats and Charles II was notorious for never leaving his spaniels, even at the Council table (Thomas 103). Attesting to their place in the domestic sphere, spaniels were England's original *poppees* (a French word that is the origin of "puppies" and referred to dogs as puppets) and lapdogs ("lapdog" was not used until the seventeenth century, making "spaniel" the principal English word for "lapdog" in the sixteenth century).[18]

Just because the spaniel or Spanish dog gained the reputation of being England's first and most endearing pet does not mean that the Spanish did not design their own version of "Spanish dogs" that has long been hidden in the reclusive corners of the field of Hispanism. Indeed, borrowing a figurative example, the dog's head was a hidden part of the Spanish domestic space. Covarrubias offers a definition of *canes* as the architectural support that holds the beams of the home: "And they call all of these *canes* because they are typically carved into the shape of the heads of dogs." Because it places the dog as a fundamental, but hidden element in the architecture of the domestic space, scholars need to better understand this hidden piece in the basic architectural foundation not just for the early modern European home, but also for the Spanish one.

Europe designed the spaniel's head to reflect their own designs for the Spaniards. In turn, the Spanish sought to develop their own version of dogs that would reflect fantasies about their own "others." In contrast to Provençe, Spain in the medieval and early modern period chose the name of dogs to disparage the outsider and ennoble itself. For instance, the name *galgo* (greyhound) represented the Spanish ennobling of self at the expense of othering France. In Spain, the hunter that owned *galgos* was considered noble. Cervantes, for instance, cracks a joke from the period that indicated that commoners had to "borrow" their *galgos* since they could not own them.[19] Indeed, the *galgo* was one of the first hunting dogs to be connected with purity of *casta.* Stevens remarks in his Spanish–English dictionary under the entry for *casta*: "a race, a breed, a progeny, a stock; 'De casta le viene al galgo ser rabilargo' it is natural, or proper to his breed for a greyhound to have a long tail" (Nieto Jiménez and Alvar Ezquerra 3: 2244).

---

[18] In the hunting treatise entitled *The Book of Albans* Juliana Berners offers the synonym *poppee* (from the French, "doll") as an alternative name for spaniel: "as a smalle ladyes' poppees that bere awaye the flees" (qtd. in Coile 2). Topsell writes that the spaniel is a "puppy" when he refers to those "pritty puppies called Spaniels gentle" (172) (that are good for assuaging ills of the stomach). The spaniel was a "pet" for petting. In 1557 one English author wrote that it was "natural" for the spaniel to be tender and loving (Oxford).

[19] In a moment of self-deprecating humor, the Grass Graduate compares himself to commoners when he states that he used *galgos* on loan: "He said hunting rabbits was quite pleasurable, even more so when one hunted with borrowed greyhounds" ("La caza de liebres dijo que era muy gustosa, más cuando se cazaba con galgos prestados" [my trans.; *Novelas* I: 57]). Also see Greer, "Diana."

Early modern Spanish writers linked the provenance of the *galgo* to France. As Sebastián de Covarrubias states: "They were given this name because they first were bred in France, or Gaul, and from that word, *galgos*" ("Diéronles este nombre, porque los primeros se criaron en Francia, dicha Gallia, y de allí galgos" [my trans.; 621]). In the same way that *espagnol* in medieval Provençe meant both "Spanish dog" and "Spaniard," so the term *galgo* in medieval Spanish meant "French dog" and "French man." The Iberians' name for the peoples to the north was the same as their word for a dog, a Gaul or *galgo* and, when Covarrubias confirmed the Gaul–*galgo* connection, he contributed to the establishment of an etymological truth constitutive of a proto-national sense of self. Moreover, just as the term "spaniel" or its semantic equivalents were not used in early modern Spain to designate a dog breed, so *galgo* or its semantic variants was not used in early modern France as a dog breed. (The French used *levrier*, a variant of *lebrel* or "rabbit dog," for greyhound). The term *galgo* was in usage in medieval Spain. But the sixteenth century was fundamental in providing the term a national connotation because humanists conceived the import of their etymological study as one that reached a closer approximation of truth. The nature of the greyhound as sign of prestige in England contrasted and compared with the Spanish. The elite, prestigious greyhound in early modern England was Irish, not French (MacInnes). The reference to the dog's origin in Ireland reflects the English attempt to domesticate and incorporate a different, disparaged regional space within the boundaries of Englishness.

Aside from the semanticization of the French into the *galgo* in Spanish lexicons like that of Covarrubias, the appearance of a number of new dog breeds in the Spanish language in the early modern period reflects the continuation of othering and ennobling with the sign of the dog. One play demonstrates how the Spanish used another dog name to other England as "Jewish." After the seventeenth-century when the English dog, especially the mastiff, had increasingly acquired fame, the word *dogo* from the English "dog" entered the Spanish language.[20] While the new Spanish dog breed of the *dogo* acted as an ennobling sign like the Irish greyhound did for the English, sources also disparage the *dogo*. An early eighteenth-century play, aware of the Benavente's seventeenth-century *entremés* entitled "The Play about the Dog's Dream" ("Comedia del sueño del perro") packages itself as Part II, entitled "The Second Play about the Dog's Dream" ("Comedia segunda del sueño del perro," 1710).[21] All the characters are animals in "The Second Play about the

[20] Although it is not found in seventeenth-century dictionaries, *dogo* was found in seventeenth-century plays like Pedro Calderón de la Barca's "Rabies" ("La rabia"). At the beginning of the eighteenth century, *dogo* increasingly appears in dictionaries. In Stevens' Spanish–English lexicon from 1706 the following entry appears: "dogo: a very large dog, as bear dog, or Irish greyhound. From the English, dog" (Nieto Jiménez and Alvar Ezquerra 5: 3840). Eighteenth-century dictionaries describe the *dogo* as a bull-baiter, synonymous with the *alano*.

[21] The play is preserved in a chapbook in the Spanish National Library and the anonymous author writes a burlesque court drama. In the title of the chapbook, the author hides his identity, drawing on the popular *romance* poetic tradition, calling himself a blind man: "Written by a blind man from the Post Office. Translated in Castilian and Portuguese

Dog's Dream" and the two protagonists, named after Cervantes' famous dog duo, Cipión and Berganza, begin the play by appearing on stage singing together. The play is filled with insipid humor such as an anti-Semitic joke in which the wolf describes the relatively new Spanish breed known as a *dogo*: "My presenter was a *dogo* / with a longer tail than a Jew" ("Mi anunciador era un Dogo, / con más rabo, que un Judío" [my trans.; 228]).[22]

Américo Castro did not consider the extent to which the animal, specifically the spaniel, played an important role in the history of Europe's conceptualizing of one of its significant "others," that is, the Spanish. Another significant omission occurs in the work of Tzvetan Todorov. In *The Conquest of America: The Question of the Other*, Todorov ostensibly argues that the Self was the Spaniard, whose possession of an instrument—most especially, language—enabled him to understand and conquer the "Other" or the Native American. In this work, Todorov humanizes the category of "America" at the expense of "Europe" and humanizes the category of "woman" at the expense of "man." In doing so, Todorov perpetuates an anthropocentric nationalist and gendering tradition that began with the European creation of the spaniel.

The construction of the human self in hierarchical relationship to the animal apparently plays a rather insignificant role in the Todorov history narrative. Although three short references to dogs appear in the amply-documented text, the dedication to his book highlights the hidden, yet overarching, role of this animal as it informs his methodology. The Spanish dog is an unacknowledged sign of othering and ennobling in the Todorov method. Before the book begins, Todorov writes a dedication in which he gives the brief account of the capture of Bacalán, a Mayan woman, who is thrown to the Spaniard's dogs. Todorov writes, "I dedicate this book to the memory of a Mayan woman devoured by dogs" (n.p.).

---

by an Armenian from the Puerta del Sol" ("Compuesta por un ciego de la estafeta. Traducida en castellano y portugués por un Armenio de la Puerta del Sol"). The play is an inside joke and commentary on members of Philip V's court. At the end of the comedy a "Key to the Meanings of the Names in the Comedy" is written in eighteenth-century script and provides the name of a court member with the corresponding animal character and dog types. Cipión and Berganza are loyal spies; the *galgos* y *lebreles* are traitors to the Crown; the *mastines* are church fathers; the *podenco* is a loyal poet, and the lion is Philip V.

22 Aside from the *dogo,* another new dog breed entered the Spanish language in the seventeenth century, the *chino.* The *chino* reflected the way in which Spain othered and ennobled itself. In the early seventeenth-century, based on the opening of trade routes in the East, the word *chino* as it referred to the Chinese was increasingly used in the Spanish language. Sailors brought back stories that the Chinese had bred small dog varieties and also brought back specimens. Soon, hairless dog varieties (especially in Peru) were referred to as *chinos* and, in the seventeenth-century, Francisco de Quevedo would write an *entremés*, "Angry People" ("Los enfadosos"), in which he makes fun of a bald man for looking like a *chino* dog. *Chino* did not just refer to the dog or a person from China, but also was influenced by the Quechua word *china* that referred to a woman, generally servant, and a female animal. *Chino* would form part of the growing logic of racial categorization and be used as a category in caste paintings, especially in the Andean region (Martínez 166).

The animality of the dog *makes* Bacalán human in that the innocent, sacrificed Bacalán is humanized into an ennobled figure, the object of the dedication of Tzvetan Todorov's book—a memorialized body contrasting that of the large dog, the iconic bestial force of the Spanish.[23]

With respect to the dedication to *The Conquest of America*, Todorov's notion of the other depends on an animal-human hierarchy in which the dog is used to establish an individual's humanity. Instead of a productive theoretical construct, this idea of the other entrenches Hispanism in anthropocentrism because it continues the logic of othering and ennobling based on the dog that began in the early modern period. Just as Europe designed a spaniel that ennobled itself and othered Spain, so Todorov uses the Spanish dog to ennoble Bacalán and other Spain.

Todorov inverts a tradition of othering through the figure of the dog that began in early modern Spain. This tradition includes: the use of the dog as a punishment tool to other Americans; the representation of new races of peoples as having canine heads; the use of canine-related terms in new human identity categories; and the Spanish parrot's use of the term "dog" for a slander. In each of these cases, as one group is "othered," so the dog is also used to ennoble, demonstrating how Todorov's portrayal of the dog simply reverses the logic based on othering and ennobling that began in force in the early modern period.

In contrast to the early modern European spaniels (Spanish dogs), early modern Spain developed its own breeds of large dogs (Spanish dogs), most especially the *mastín* or Spanish mastiff, the *alano*—the first dog breed in Europe to be celebrated in verse, and the *galgo*—Spain's version of the greyhound. As opposed to the popularity of the word spaniel across Europe, spaniel *never* entered the Spanish language in the early modern period. No Spanish dog was ever called a spaniel or any semantic variation thereof as the Spaniards simply used the word *gozque* for the small dog (see Chapter 3). As opposed to Europe's "Spanish dog" or spaniel, Spain designed its own dog as large, powerfully masculine, and a tool of war. By the seventeenth century, the fame of the Spanish dog in the conquest was a commonplace in places like France and England. As John Florio writes in his 1603 English translation of Michel de Montaigne's 1580 essay "An Apologie of Raymond Sebond:"

> As the Spaniards did to their dogges in their new conquest of the Indies, to whom they gave wages and imparted their booties, which beasts shewed as

[23] One can find examples of ennobling and othering of the dog in visual representations of Mexican history. One notable example can be found in the 1941 mural *Colonization* found in Mexico City's National Palace. Diego Rivera ennobles the indigenous Mexican by making the hairless xolo a symbol of national pride in contrast to the othering of the imperial Spaniard. In the mural, Rivera places a xolo on the right at a noble Aztec's feet placed on a platform above a fierce, barking Spanish dog who accompanies an animal-like Hernán Cortés, green-faced and microcephalic (Lerner). Xolos are not only protagonists in some of Rivera's paintings as well as Frida Kahlo's, but they also both had xolos as pets. Xolos frolicked in their Blue House in Coyoacán.

> much dexteritie in pursuing and judgement in staying their victorie, in charging or retreating, and, as occasion served, in distinguishing their friends from their enemies, as they did in earnestnesse and eagernes. (179)

Just as the popularity of the spaniel surged, so the fame of the Spanish soldier dogs increased across Europe, especially in many texts that sought to perpetuate the *leyenda negra* (Spain's unjust brutalization of conquered peoples). In their construction of their English version of the Spanish dog, the spaniel, and their version of the true English dog, the mastiff, the English *followed* rather than set the precedent for the Spanish large dog with respect to the symbolic and material imperial design.

In the years that they designed the small Spanish dog, the English adopted Spanish conceptions of the large dog into a unique English breed, the mastiff. In Spain, in one poem the *alano* was the noblest of all the dogs: it was "León"—the "lion of dogs." In turn, in England it would be written in the eighteenth century that "what the lion is to the cat, the mastiff is to the dog, the noblest of the family" (qtd. in MacInnes 31). Like the large Spanish dog, that had been set upon the Americans in the sixteenth century, so mastiffs were set upon North Americans in the seventeenth- and eighteenth-century. Although not employed as extensively as Spanish conquest canines, in Plymouth Harbor in 1603, two English mastiffs named Bristol Foole and Gallant accompanied the Englishman Martin Pring. The Indians were more afraid of these two dogs than twenty of Pring's men. One of the mastiffs carried a pike in his mouth (see Mastromarino for more cases of English soldier dogs in the colonies).

Borrowing an established Roman tradition, the Spanish dog was a soldier and regularly used to punish Native Americans one hundred years prior to the English use of its own soldier dogs.[24] Resulting in a specifically American-based historic trauma based on the Spanish dog and influencing the Todorov theorization of the other, Spain developed the dog punishment into a full-fledged practice in the Americas over the next three centuries, a practice that largely disappeared in the medieval period in Europe. The practice of dogging or *apperramientos* incorporated a penal mindset in which the punishment itself fit the crime and criminal.[25] The medieval logic of the dog-on-dog type punishment (the real dog punishes the human "dog") flourished for centuries after the conquest in the

---

[24] Pliny's *Natural History* (translated by two royal doctors, Francisco Hernández for Philip II and Jerónimo de Huerta for Philip IV) provides a historical precedent for canine soldiers (Cantó). The dogs were the most loyal of the troops, Pliny writes, and they did not demand a salary (Diel).

[25] As they utilized dog soldiers, a common Roman practice, so the Spanish also revived the Roman practice of using dogs as punishment tools. The Roman logic consisted of making the punishment fit the nature of the crime and criminal. One medieval example of this practice was punishing Jews by hanging them between two dogs in an inverted parody of the crucifixion (Boehrer, *Shakespeare* 26; also see Kalof 60–61). Various scholars have studied cases of animals that are punished like humans (see especially Evans, Thomas 97). More study is needed on the punishment of animals in the medieval and early modern Spanish context.

Americas.[26] Using the newly-conquered space of the Americas, the sixteenth-century Spanish reverse the terms of the way in which Europe constructed the Spanish dog. The Spanish humanize themselves at the expense of animalizing the American. The Spanish associate themselves with the dog by molding the canine into a punishment tool that incorporates the attributes of an ideal soldier, including powerfully-contained violence, military prowess, masculinity, and nobility. In turn, the Spanish paradoxically disassociate themselves from the dog since those deserving of punishment are also canine: animal-like, feminine and weak.

The simultaneous association and disassociation with the dog in the Spanish context connects to the emergence of the word "Spaniard" that took its modern shape in the sixteenth century when outside geo-political influences from Europe and America inspired the formation of the national human category "Spaniard." The dog was a significant sign of othering in Spain's growing imperial designs, particularly related to the head of the dog. Spain's early explorers imagined dog-headed races as living in the outer reaches of the known world, living on the borders between the savage and the civilized, between the animal grunt and human language (Vignolo). Marco Polo had reported sightings of cynocephali on the Andaman Islands of the Bay of Bengal and one of Spain's first conquests, the Canary or, literally, "Dog-Men" Islands, was named after the legend in which the borders of the world harbored this fantastic race of Dog-Men (White 59; 61–70). This process of othering the inhabitants of the conquered territory with a canine sign also connects to the ennobling act of appropriating the dog as self. Spain othered the region through the sign of non-human dog-headed peoples, but ennobled itself through the act of conquest and naming the new land after the dog, specifically, the Canary Islands. Indeed, the Spanish put on a dog face with pride when they placed the emblem of the dog on the coat of the arms of the island.[27]

The othering-ignobling process through the sign of the dog continued in the islands occupied by the Spanish in the Caribbean. The name of Spain's newly-conquered territory "Hispaniola" (literally, "La española" or "Spanish island" in which the word "island" is elided) was written on the first European maps of

[26] Except for in the colonial Americas, the punishment by dogging generally ended in Europe after the Middle Ages. For use of dog punishment in a variety colonial Spanish contexts, see Varner and Varner; Diel, Valle. For use of dog punishment against blacks in Cuba and linking with events during the Haitian Revolution (1791–1803) and the Second Maroon War in Jamaica (1795–1796), see Johnson. For dogs against natives in the Second Seminole War in territorial Florida (1835–1842), see Johnson.

[27] In a manuscript from about 1580 (Biblioteca del Real Monasterio de San Lorenzo del Escorial, sign. 28–I–10, 11 y 12, vitrinas 21, 22 y 23) entitled *Nobility of Spain* (*Nobiliaro de España*) by Alonso Téllez Meneses, a dog with a golden collar is used for the representation of the Kingdom of the Canary Islands ("El escudo"). Later, in the first printed edition of the *History of the Canary Islands* (1772) the frontispiece includes a coat of arms with two collared dogs holding up an image of the seven Canary Islands by José de Viera y Clavijo, dogs being the etymology of the name of the island (Erbez). A version of this coat of arms is used today.

the newly-drawn Americas.[28] The canine also haunted the ground of the name of the new American territory since, for Europeans, *La* Española resonated with a gendering of the New World and also with the canine because the new name for the Spanish island also meant "Spanish bitch," the same European word for a female Spanish dog. Neither Columbus nor the Spanish people consciously thought of the canine version of *hispaniola* (female spaniel) in naming the new space. The dog, however, was present in the semantic connotations of the name. The dog was even more transparent in the naming of the Canary Islands and other naming acts at the outset of the Spanish discovery. Columbus, influenced by Marco Polo and the medieval legends of dog races, describes the inhabitants of the islands as cynocephali in early letters. The name "Carribean" also enclosed the sign of dog. When he named the place "Cariba," Columbus confused it with "cani-ba", a term meaning the canine Khan (White 59). The early word for the people from the Carribean region, cannibal, also connected those supposed human-flesh eaters with the dog because it contained the *can,* the Latin word for dog.

The dog is hidden to different degrees within the names of the "Spanish" themselves as well as the Spanish colonized spaces. Ever since Spain's arrival to the Americas, Europe used the symbolic weight of the dog in a process of othering Spain based on region, race, nation and gender that also ennobled itself. In turn, the generic terms used to classify animals and humans (those terms related to *casta*) took on new meanings related to race in the context of the sixteenth-century Hispanic Atlantic. The American experience radically influenced the semantic charge to the notion of *casta*. Spain quite literally animalized its others as the word *casta* would provide a zoological inflection to Americans in the newly created *casta* books in the sixteenth century and later the *casta* paintings. Words like Mestizo and Cholo originally meant "dogs." *Mestizo* was first used to define a mixed breed canine and *cholo* was a Native American word that meant dog. Both words labeled different types of peoples and can be found in caste paintings and many other sources.[29] The genealogy of each word elucidates how modern identity categories are built upon the disassociation and association from the animal since *mestizo* and *cholo* were initially pejorative when they originally referred to people (they were "dogs"), but each formed part of a larger trajectory of a dual process of othering *and* ennobling since they were also to soon become "humanizing" words in the sense that each word also would connote dignity and pride for a specific American identity group. This process of othering and ennobling continues to inform criticism up through Todorov. (See the appendix for more detail on the etymology of Cholo and Mestizo, as well as other terms—such as Mulatto, Creole,

[28] Map makers adopted Columbus' naming of his newly-discovered island, "La Española," into "Hispaniola" based on Peter d'Angiera's Latin treatise that appeared shortly after the Spanish colonization of the Caribbean.

[29] For one reading of violence associated with animalization in caste paintings, see Guzauskyte.

Maroon, and Mustang. These terms are first zoological terms that often disparage people, but then revert to ennobling categories.)

The dog was not only used to other and ennoble in the context of the Americas, but also in Spain and Europe. In fact, the body of the parrot, an animal that was brought back from the Americas, reflects another way in which the Spanish othered and ennobled.[30] The Spanish parrot, another one of the earliest symbols that Europe used to represent the Americas on maps, often spoke and, each time it spoke, it called the "Moor" a dog. In contemporary English, the parrot is often portrayed as saying "Polly wants a cracker," but in early modern Spain the parrot's words were not as anodyne as the Polly expression. The Spanish parrot ventriloquized a dog slander against the Moors. Its proverbial expression—found in literature from the early sixteenth through the seventeenth century—was *perra mora* or *perro moro*, expressions that roughly mean "Muslim bitch" or "Muslim dog."[31]

The parrot was kept within the intimate domestic confines like the dog and performs an act of othering by incessantly repeating a racial and religious canine slur.[32] The expression "Moorish dog" a religious slander followed medieval precedents in which those that practice the outsider religion, Islam, were called

---

[30] Gonzalo Fernández de Oviedo writes in his *General and Natural History of the Indies* that: "so many species [of parrot] have been carried to Spain, it is hardly worth while to take time to describe them here" (qtd. in Boehrer, *Animal* 80–81).

[31] One finds the parrot expression *perro moro* in Juan del Encina's early sixteenth-century "Interlocutory Eclogue" ("Egloga interlocutoria"); a late sixteenth-century poem by Baltasar de Alcázar (190); and the mid-seventeenth-century *mojiganga* by Vicente Suárez de Deza (qtd. in Borrego Gutiérrez 187–88). Lope de Vega's play *If Women Could Not See* (*Si no vieran las mujeres*) provides another example. A character in the play tells the anecdote of a peasant who listened to a talking parrot carrying on an animated conversation with a lady at the window. The peasant then asked the parrot: "How are you? It replied '*perro moro*' / in his half American accent" ("¿Cómo estás? y él, 'perro moro' / con su media lengua indiana" [my trans.; Vega]).

[32] The target of the parrot's *perro moro* expression is simultaneously religious and racial. "Moor" referred to both those who practiced Islam as well as those who had darker skin color. The seventeenth-century play, *The Valient Black in Flanders* (*El valiente negro en Flandes*), begins with the expression *perro moro* when one character calls the black protagonist Juan a dog. The black protagonist Juan responds to his antagonist by saying that *perro moro* applies to Moors and not to black men like himself. In passing off the slander onto the Moors, Juan attempts to dignify himself by playing the role of a "noble" dog (Beusterien, *An Eye* 117). But Juan's plea to not be considered a "Moor" would have not been wholly persuasive to his audience. The phrase *perro moro* was directed against Muslims, but it was also, in the popular imagination, directed at those of a darker complexion. A common expression from the period indicates that Moors were classified as dogs just like slaves: "We call Moors and slaves 'dogs' because no one saves their soul and they die like dogs" ("Perros llamamos a los moros y esclavos porque no tienen a quien les salve el alma y mueren como perros" [my trans.; Correas 629]). The popular expression disparages both religion and race because black slaves were not necessarily Muslim. It was common practice to baptize all black slaves in early modern Spain.

"dogs." But it was also racial in the sense that in the Americas those that were called "Moors" were often considered those of a mixed race heritage even though they may very well have never practiced or even heard of Islam. Indeed, the meaning of "Moor" as not so much as one who practiced Islam, but classified as a racially-impure individual heightened in the Iberian Peninsula and Iberian world after the conquest of the Americas.[33] Therefore, the parrot othered by using a canine slur that was symptomatic of an early modern Spanish racial and religious slandering ideology. The parrot, however, also ennobled since those who owned it had the special privilege of owning a luxury item. Parrot ownership formed part of an ennobling prestige—a growing way to enhance one's "noble status" in a similar to the way in which owning a spaniel gave prestige to the English family prestige or owning a *galgo* gave prestige to the Spanish family.[34] The parrot as luxury item, apparently "just" an animal within one's domestic space, was much more since its body conjured up the symbolic space of the Americas as well as symbolically heightening its owners' "humanity."[35] In short, a bird grants prestige and humanity by animalizing Spain's others.

An un-articulated and un-theorized animal, especially the dog, can be found in the context of early modern Spain. In the context of Castro's work the spaniel is not an important term in historicization of "Spaniard." Todorov does not study dogs, but bestows upon the Spanish dog an overarching role in establishing the humanity of the other. Through the tracing of the genealogy of the dog as a significant sign in the words spaniel and *español* and in Spain's American experience, the critic may better recognize how divisive regionalist thinking based on an inside-outside anthropocentric mentality emerged in Hispanism. Ever since the parrot's first arrival to Spain from the Americas, the sign of the animal formed part of this inside-outside antagonism.

Scholars are starting to grasp the nature of how the material sign of the animal has been hidden and one recent study on Latin American history has called for the importance of studying the "real" animal (Few and Tortorici). The intense affect that people have for their domestic animals, especially the dog, is one reason that the sign of the animal has been hidden away in the house that constitutes

33 "Morisco" was not religious, but racial in the Americas (see Beusterien 38 and Martínez 165).

34 After Columbus brought back colorful parrots as gifts for Isabella and Ferdinand, American parrots were in European markets, such as in Seville, fetching high prices as a luxury commodity (Boeher, *Parrot*).

35 The sign of the parrot was a symbolic appropriation of the American space. Since classical writers had described parrots as a bird from India, the parrot was a confirming presence that Europe had "truly" arrived to the East (Boehrer, *Parrot*). Within a decade of Columbus' landing, chroniclers had mapped out the newly-discovered territory and, in a cartographic act of appropriation, images of the parrot frequently mark the space of America itself by appearing as a symbol of America on early maps (George). Later, Philip II ordered the painting of American parrots on the ceiling of his private chambers in the Escorial, many directly based on the Hernández expedition to America (de Asúa and French 93–4).

the field of Hispanism. Tied to the economy of the new global system, animal affect uniquely emerged in the early modern world—in Europe *and* in Spain—and is evidenced by the rise of pet culture. With the growth of towns, a notion of the interiority of the domestic space, and increased leisure time, peoples in early modern Europe increasingly developed sensibilities to animals that formally were viewed unsentimentally (Thomas 102, 110; see also Shell, "The Family;" Bohrer, "Shylock;" Tortorici). By studying the case of early modern Spain within the wider geographical scope of the Atlantic world, Europe, and especially England, critics may better recognize how anthropocentrism in Spain functioned in a similar way in a European context that both influenced Spain and was influenced by Spain.

Rising animal affect surreptitiously masked an emerging nationalism in the formation of self in early modern Europe (one may draw a contemporary parallel by pointing to the link between the Chihuahua dog in the private space of the home and its apparent irrelevance to the Mexican state of Chihuahua). The affect for the spaniel plays a role in hiding formational nationalism and gender since as sixteenth-century English people sought the company of spaniels, they did not think "Spaniard," just as the sixteenth-century Spanish nobleman did not think "Frenchman" when he called on his *galgo.* The "forgetting" or obfuscation of the national other in the animal's name is due to the ease in which the animal, most especially, the dog becomes naturalized as part of human affect and as part of the domestic space. During the early nineteenth century, the logic of the *galgo* returns to Spain when the French were animalized as dogs once again. Soult and Ney, the names of two marshals from the invading French armies, were used for typical household dog names. Today, in the Galician countryside versions of the dog names Soult and Ney are common, but their owners are unfamiliar with the history of their dogs' namesakes (Rivas).[36]

A complex reciprocal process of ennobling and othering occurred in early modern Europe in the naming of dog breeds. England and other countries adopted a version of the Spanish dog and Spain, in turn, adopted its own breeds, influenced by and influencing the dog design practice in early modern Europe. Moreover, Spain, beset by canine thinking, associated and disassociated itself with the dog in naming and conceiving its imperial space. The canine naming process is an entrenching process that congeals a notion of the human through the rhetoric of a disparaged outsider other versus a dignified inside subject and this practice of canine ennobling and othering continues from the early modern period up through Todorov's work.

For this reason, it is beneficial that Hispanism look outside of the language of the animal as "other" and to consider it an "other other." Because of the emergence of the early modern European domestic space, the period is especially important for discovering the design of this other other. When Juliana Schiesari looks to

[36] There is a national park in Spain that is called "Despeñaperros" (Dog Precipice) and the origin of the place name is generally thought to be that the Spanish threw the nineteenth-century invading French forces or the "dogs" off the cliff to their death.

the reclusive corners of the domestic world of Italian Renaissance, she finds "the beginnings of today's 'post-humanist' critique of humanism and the posthumanist interest in nonhuman others" (*Beasts* 6). One may read Covarrubias' assertion about the dog's head as a hidden part of the Spanish domestic space as a way to place the dog as a fundamental, but hidden element in the architecture of the domestic space. In this way, Covarrubias' definition of the dog unveils and illuminates the broader assertion of this chapter—that the animal, most especially the dog, is a hidden piece in the basic architectural foundation for the home that many call the field of Hispanism. A close study of the animal, especially the dog, reveals how this canine architectural support played an important role in perpetuating anthropocentrism in certain aspects of critical inquiry on early modern Spain in the twentieth century.

This chapter has suggested that a better recognition of the sign of the animal in the architecture of the field of Hispanism provides a productive blueprint for future practice in the field: a path for theoretical practice that thinks beyond anthropocentrism and the rhetoric of the inside versus outside. Thinking ahead to the next chapters, this chapter has set the groundwork for a dialogue between Animal Studies and Hispanism that will better articulate the "other" other, looking beyond the perniciousness of anthropocentrism, so that the field can *be*—neither on the inside nor the outside—operating with a theory and praxis that are not imprisoned and immobile.

Chapter 2
A Cervantine Animal Exemplum: Animal Studies and "The Dialogue of the Dogs"

> Cipión: But stay with this, since I don't want us to seem like some know-it-all preachers… (Mas quédese aquí esto, no quiero que parezcamos predicadores… [37; 311]).

Rather than studying the animal itself, Sigmund Freud often uses the metaphor of the animal to describe the workings of the human subconscious. Famous Freudian cases such as the Wolf Man contributed to the construction of a notion of a human psyche constituted by an "animal other" that inhabits the unconscious, our most familiar and unfamiliar territory. One Freudian origin myth perpetuates the notion of animality as part of the human psyche. It tells the story of when humans separated themselves from the animal. At an early period the human was sexually excited by the smells of feces and blood. But, at "some" point in time, the human found those smells disgusting and repressed them. The organic repression of feces and blood would lead to a human culture obsessed with cleanliness that privileged sight over the olfactory and would lead to the sexual repression that enables the founding of the family and civilization. At the critical moment in which feces and blood became objects of repulsion rather than stimulation, humans became *real* humans and not animals.

Animal Studies scholars have re-written long-assumed human origin theories based on the animal. Cary Wolfe signals Freud's creation of an empty genesis moment of the human break from animality (*Zoontologies* 1). Freud assumes the origin of the human is at the moment when it decides that the organic is disgusting and must be repressed. In fact, in *Civilization and its Discontents,* he writes that dogs were reprehensible because they had no horror of excrement and not ashamed of their sexual functions (4495). His assumption, however, is based on the fact that the human must have known—before becoming human—that the organic should be repressed. Freud, then, creates the notion of the temporal with respect to the question of the emergence of the human psyche as out of and after the animal. The temporal break forward from the animal is established as fact, but a fact whose logic is fallacious since the moment of change exists at a fabricated past mythic moment.

Dogs were beloved companions for Freud until the end of his life (Molnar). Despite his association between the canine and feces and sex, the dog itself was not the object of his studies. Another way that Freud did not let the dog be a dog was when he read Miguel de Cervantes' "The Dialogue of the Dogs" ("El coloquio de los perros"). Instead of understanding the dog as animal, Freud himself playfully adopted the role of Cipión from the dialogue. Freud first read "The Dialogue of the Dogs" in the textbook that he was using to learn Spanish and he was markedly influenced by it at an early age. In a letter to Martha Bernays from 1884, Freud writes the following about Silberstein, an old friend from school:

> We learned Spanish together, had our own mythology and secret names, which we took from some dialogue of the great Cervantes. Once in our Spanish primer we found a humorous-philosophical conversation between two dogs which lie peacefully at the door of a hospital, and appropriated their names; in writing as well as in conversation he was known as Berganza, I as Cipión. How often have I written: "Querido Berganza!" and signed myself "Tu fidel Cipio, pero en el Hospital de Sevilla" (sic). Together we founded a strange scholarly society, the "Academia Castellana," compiled a great mass of humourous work which must still exist somewhere in my old papers. (qtd. in Riley 1)

Freud and his friend Silberstien's adoption of the dog-personalities from Cervantes was not a passing, forgotten game. Throughout his life—as late as 1928, 44 years later!— Freud would remember when Silberstein and he adopted the roles of Cipión and Berganza (Riley 1).

The following chapter turns away from interpretations of the canine in the "The Dialogue of the Dogs" as a figure or mask for the human, such as Freud's adopted role of Cipión. Instead, it turns to Animal Studies in order to argue on behalf of the elimination of the animal as figure to better conceive the interdependency of animal and human.[1] The chapter first reads the dog dialogue as reflective of Cary Wolfe's critique of theories that assume that language belongs to humans. Then, through a close reading of Berganza's adventures, the chapter looks to ways in which Cervantes ingeniously creates a new form of animal exemplary literature that muddles the animal's role as a positive and negative model for human behavior. It pays special attention to the first episode told by Berganza from when he lived at a butcher shop. The episode highlights the dog as a model for exemplary behavior by criticizing the butcher's lasciviousness, by abandoning its role as companion to the bullfighter, and by condemning the butcher's murderous ways. The second episode—Berganza as sheepdog—also highlights the dog as a model for exemplary behavior in contrast to deceitful wolf-like shepherds. "The Dialogue of the Dogs," however, is not a traditional animal exemplary tale. In later episodes, Berganza is portrayed a model of evil behavior, especially in his treatment of an Afro-Hispanic woman and the witch, Cañizares. After confounding the two typical models of exemplary literature—animal as exemplary and animal as a model of vice—Cervantes presents a new animal exemplary episode in which animals learn from humans how to behave.

---

[1] Rather than studying the animal and human interdependency, Abel Alves interprets the dog dialogue as one that breaks down the species boundary because the dogs work as figures for what goes on in human society. He focuses on the fact that the dialogue is a literary anthropology of Cervantes' society (57).

Ever since the classical period, the ownership of true language was reserved exclusively to humans. When the sixteenth-century humanist called on the dog, it did not speak, but, lying asleep by his side, inspired him as he worked tirelessly in his truth-finding pursuits. The dog, as simile, was also called on to retrieve creative ideas. To characterize their work, humanists used the simile of the action of the hunting dog. The dog's ability to find its quarry and retrieve it for the hunter was an extended simile to describe the creative labor at finding hidden meanings in language and for successfully communicating those findings.

The author of *La Celestina* begins the opening pages of the work with "The Author to a Friend" ("El autor a un su amigo"), stating that that he is trying to figure out a way to repay his friend's generosity. Placing himself in the posture of the thinking humanist, he explains that he sits and deeply meditates on how he should repay his friend. His mind goes on the hunt like the scent dog: "I have on, various occasions, retired to my chamber and there meditated, with cheeks resting in my hands, my senses ranging out like hunting dogs, my judgment soaring" ("asaz vezes retraido en mi cámara, acostado sobre mi propia mano, echando mis sentidos por ventores y mi juicio a volar") [3; 184]). In the following sentence, the author implies that the dogs brought him a quarry, that is, the creative fruits of his meditation, *La Celestina*: "And then it seemed very clear to me that our native land, because it has so great a number of gallants and lovers, has much need of this present work" ("me venía a la memoria, no sólo la necesidad que nuestra común patria tiene de la presente obra or la muchedumbre de galanes y enamorados mancebos que posee" [3; 184]). The hunting dogs retrieve the idea for the book itself since they, raging out like his mind, bring the creative material that will become the written text of *La Celestina*, a text that will help his friend.

Juan Huarte's *The Examination of Men's Wits* (*Examen de ingenios para las ciencias*) repeats the hunting dog simile with respect to its connection to creativity—in this case for the orator. For Huarte, the exemplary orator must possess an imagination like a hunting dog since he has "a heightened imaginative faculty like a scent dog that looks for and retrieves the quarry to your hand" ("muy subida imaginativa que sea como perro ventor que le busque y traiga la caza a la mano") [my trans.; 439]). When he creates his talking dogs, Cervantes eliminates the humanist canine simile. Berganza does not call on the figure of the hunting dog in order to inspire himself to tell the story. Berganza and Cipión do not look to the simile of the hunting dog as they craft their eloquent speeches. Berganza *is* the story-teller. Berganza and Cipión *are* the skilled orators—replete with wit and imagination.

The classical humanist tradition connected the animal with the lack of language. As Prospero, a representative of the humanist, says in Shakespeare's *Tempest*, "brutes did not speak language, but only gabble." The animals in important source texts of "The Dialogue of the Dogs" are bereft of language. The ass in *The Golden Ass* is mute. The stag is also mute in Ovid's *Metamorphosis*. After Diana turns Acteon into a stag, Acteon realizes he has no language. Ovid writes: "But when he sees his head and horns reflected for certain in the water he tries to say 'oh, look at me!' But no voice follows" (Ovid). In the renaissance humanist tradition that revives Ovid's *Metamorphosis*, language comes from a human tongue. A dog dialogue was published in French some 80 years before Cervantes' publication of his own dog dialogue. The Lucian-inspired Bonaventure des Périers makes his two talking dogs, named Barker (Hylactor) and Glutton (Pamphagus), two of Acteon's dogs who mistook him for the stag and ate him. In his dialogue the dogs eat a human tongue—Acteon's—and can have their conversation. Speech in the dialogue therefore depends on the dog's incorporation of the human tongue. The tongue acts as a broader metaphor for animal talking in general in fables and other texts since, if animals spoke in the animal fable and exemplum tradition, they were mouthpieces for human speech.

One of the most basic ways that early modern humanism expressed human exceptionalism was through the assumption that only the human speaks and the closer the human could connect to true language, the closer it reached divine truths. One of the main projects of the humanist—based on this human-based linguistic premise—was the study of language and the recovery of classical sacred languages such as Latin, Aramaic and Greek. The humanist did not grant the animal the power to communicate truths, but sought the yet-to-be-discovered ideal human language that mirrored divine speech. For instance, an ideal tongue was bestowed on humans by God, impelling the study of ancient languages in order to uncover the vestiges of the ideal divine tongue (Forcione 223). In his reading of the dog dialogue, Alban Forcione directly contrasts the human ability to speak with animality. Forcione writes that Cipión's pronouncement that "I'd hate for the sun to light up the sky and throw us back into a dark age of silence" ("al salir del sol quedásemos a la sombra del silencio" [69; 333]) means that when the dogs no longer have the power to speak, they will return to the "dark silence of animality" (*Cervantes and the Mystery* 221).

But, at the beginning and end of the dialogue, when the dogs extol the wondrous benefits of language, the praise is never delineated as "human." In contrast to the Ovid story, Berganza and Cipión—whether they be humans metamorphosized into animals or just animals—have the power to speak on one special night. Cipión and Berganza are not mute, like the stag in Ovid or ass in *The Golden Ass*. In contrast to Périers' dog dialogue and other anthropocentric recuperations of the Acteon story, Cervantes never mentions that the dogs have consumed or incorporated a human tongue in order to speak (even though they demonstrate superb linguistic wit and amazing poetically-charged rhetorical turns). In short, while he praises language, Cervantes does not espouse *human* exceptionalism with respect to language.

Cervantes does not clarify whether the dogs are animal or human, thereby leaving open the matter of the "dark age of silence." Berganza begins the dialogue by telling Cipión that he believes he was born as a dog at the butcher's shop, but his experience with the witch Cañizares puts that origin into question. She tells him that he and another dog are the sons of another witch, Montiela, whose two children were transformed into dogs by another witch, La Camacha. This intentional lack of clarification means that from a narrative point of view, the dog's ontological status is irrelevant in the consideration of language. Indeed, since the origin of the dogs is a central unanswered question, Cervantes deliberately emphasizes the murkiness of the animal-human linguistic quandary. In leaving open the question that language necessarily belongs to the human, Cervantes opens the possibility that human control over language is unfounded and not "ours" at all.[2]

Aside from the animal-to-human chronology in the Freudian origin story, Cary Wolfe also calls into question a basic chronological assumption in humanist theories about the emergence of human language. Wolfe argues that human language did not appear sometime *a posteriori* to animal communication because scientists have shown the existence of a shared animal and human language that provides evidence that linguistic domains emerged *before* humans. Language already existed in processes of social life and communication among animals and no truly novel traits have been proven to exist in human language to support human uniqueness from animals. For Wolfe, the human does not only make language its object, but it is also the object of language in the sense that language is part of an a-human evolutionary process that makes the human its subject. Language should be recognized as the mode of consciousness associated with it, but is also tied to an evolutionary substrate. As Wolfe writes:

> Rather, "we" are always radically other, already in- or a-human in our very being—not just in the evolutionary, biological, and zoological fact of our physical vulnerability and mortality, which we share, as animals, with animals, but also in our subjection to and constitution in the materiality and technicity of a language that is always on the scene before we are, as a precondition of our very subjectivity, for what makes us human. ("Flesh" 28; "Human" 571)

Wolfe cites the case of animal language with the example of the dog Rico who is capable of fast mapping, an ability to instantly assign a meaning to a new word, a fact that leads him to cite a study by Noam Chomsky and Marc Hauser that rejects human exceptionalism when it comes to language. Sharing with many animals the intraparietal cortex (an area of the brain in which a language system is based), the human language system is shared in evolutional and biological terms with many other species ("Thinking").

Beyond raising the question of the human and its connection with language, Cervantes uses the dog dialogue to craft a revolutionary animal exemplum. Animal exemplum literature—an extremely popular genre in Europe from the medieval

[2] For this reason, I have given preference to the stuttering thesis as an influence in Cervantes' creation of the talking dogs, as well as an influence on his literary creation. Stuttering is a shared human-animal trait and, I believe, is one important material linguistic phenomenon that was on the scene before we were. I think Cervantes' stuttering inspired him to conceive the human-animal divide in the innovative ways that he does (Beusterien, "Did"). For animals as stutterers, especially songbirds, see Shell, "Animals."

through the early modern period—functioned in two ways.[3] The animal served as a model for either exemplary or evil behavior. Through the theory of mimesis, Aristotle conceived of literature as a way in which humans learn right and wrong through imitation. In animal exempla, when the animal acted in an ethical and just manner, it functioned as a model for humans to emulate. When it acted malevolently, it enabled people to recognize evil and thereby avoid malicious and immoral behavior. In both cases, the animal is a figure; it is a human mouthpiece for aiding one to learn how to act ethically. Furthermore, in Aristotelian logic the reverse could never happen, that is, the animal could never learn from the human since only the human—not the animal—could reason and thereby learn right and wrong through imitation.[4] Hence, in the two modes of animal exempla, the animal either acts as antagonist, a negative force from which humans learn to avoid evil, or, when it is a protagonist, to aid one to learn how to act ethically.[5] While he references both animal exemplum traditions, Cervantes ultimately constructs a ground-breaking version of the genre.

"The Dialogue of the Dogs" is between two *alanos*. The *alano* was a medieval hunting breed and became the most distinguished Spanish breed in the early modern period.[6] It consists of one of these dogs, Berganza, telling his life story to Cipión, who often interrupts Berganza's storytelling with comments and questions. Following the structure of the picaresque novel, Berganza recounts his life as a succession of experiences with different *amos,* a word that can mean "employer" and "owner." Berganza tells Cipión that he has worked as a bullbaiter for an evil butcher, as a sheep dog for a group of deceiving shepherds, as a family guard dog for a merchant, as a police dog for a constable, as companion to soldiers with whom he performs as a show dog, as a witch's companion, as a gypsy's dog, as

---

[3] For an overview of the popularity of animal exempla in medieval Europe, see Salisbury 81–107. For Aesop in Spain, see Carranza.

[4] In Aristotle's *History of Animals,* he does write that the elephant "both learns and understands" (630b18–30). But ultimately he assigns the inorganic rational soul to humanity alone: "The only animal which is deliberative is man. Many animals have the power of memory and can be trained; but the only which can recall past events at will is man" (488b24–30). See Boehrer, *Animal Characters* 15–16.

[5] Bautista Capataz's 1612 commentary of Cervantes exemplary novels connects them to Aristotelian thinking, suggesting that readers would be inspired to imitate good or avoid bad behavior. Bautista Capataz concludes: "they teach with examples when to flee vice and follow virtue" ("enseñan con sus ejemplos a huir vicios y seguir virtudes") (my trans.; qtd. in Wardropper 153). The argument in this chapter refutes Capataz to the degree that the figure of Berganza muddles the mimetic principal of knowing when to "flee vice and follow virtue."

[6] For historic references to the *alano* breed prior to the early modern period such as in the work of the medieval writer Gonzalo de Berceo, see Río Yenes and Parra Gordillo. For the most complete medieval description of what seems to be a proto-version of the breed, see Alfonso XI's hunting manual (Alfonso XI 48–50).

companion to a Morisco farmer, as companion to a starving writer and dramatist, and, finally, as companion to Mahudes, a monk who collects alms in a hospital.

Cervantes uses Berganza's first experience at a butcher shop to clearly signal the dog as exemplary figure. One sixteenth-century Spanish source text highlights the extraordinary virtue of an amazing dog that carried meat in a basket. Recorded by Luis Pérez in his treatise on horses and dogs addressed to Philip II, the dog story goes as follows:

> Your majesty, should not the following animal story be known by you and everyone? … They sent this dog to the butcher's with a basket in his mouth to pick up meat. They would put two or three *reales* in the basket or the appropriate amount and the dog arrived at the butcher's with the other customers. The butcher knew the dog and he took the money that he carried in the basket and gave the dog the amount of meat that corresponded to that amount. The dog returned with his basket full of meat and he knocked the doorknocker with his paws as he normally did and handed over the meat without any of it missing. (my trans.)
>
> Vea vuestra majestad si es este caso de animal para dejarse de decir a vuestra majestad y a todo el mundo … le enviaban a la carnecería con la cesta en la boca a traer carne: y metíanle en ella dos o tres reales, o lo que les parecía y llegábase al carnicero entre la gente. Y el cortador que ya le conocía, le tomaba los dineros que llevaba en la cesta y le echaba toda la carne que montaba los dineros y el can se volvía con su cesta llena de carne a su casa y tocaba con las manos del aldaba como solía y daba su carne sin faltar cosa alguna.[7] (26–7)

In contrast to this account in the source text, in Cervantes' version of the story, the dog belongs to the butcher and carries a basket of meat to his lover. Nonetheless, the spectacularity of the canine feat is the same in Cervantes and Pérez's story: the exemplary animal refrains from eating meat and carries it from one person to another in a basket.

Indeed, another popular version of the story uses the dog for the opposite purpose: the animal operates as model of avaricious behavior. Aesop's fables were quite popular in Spain in the early modern period and some of his fables can be found in the influential Spanish medieval text, *The Book of Good Love* (*El libro de Buen Amor*). One fable, entitled the "Exemplum of the *Alano* that Carried the Piece of Meat in its Mouth" ("Enxiemplo del alano que llevava la pieça de carne en la boca"), describes a dog that carries a piece of meat ("Enxiemplo"). It then sees its reflection in a river and thinks that it is a different dog with another piece of meat. Because of its greed and gluttony for the meat that it thinks it sees in the reflection, it drops the piece of meat. Recalling this source text, Berganza's exemplarity is further highlighted because he does what the dog did not do in the fable: Berganza is an *alano* and refrains from eating the meat.

[7] This same passage can also be found in *Del can y el cavallo* (Paris, 1583) by Julián de Medrano.

Another aspect of Berganza's exemplarity is his carnal self-control in the sense that he avoids lasciviousness. Berganza prides himself on his honorable behavior when face to face with a fair and beautiful girl. Berganza brings the basket of meat to his master's lover and she hands over her *chapín* or shoe in exchange for the contents. The lust ridden butcher exchanges his "meat" for his girlfriend's *chapín*, a special elevated shoe worn by women that many authors linked to women's sexuality. In contrast to the libidinous butcher, Berganza not only does not eat the meat in the basket, but states: "I didn't want to sully those clean white hands with my dirty, bloodstained mouth" ("no quise, por no poner mi boca jifera y sucia en aquellas manos limpias y blancas" [30; 305]). The butcher has the telltale signs of sexual disease on his face; his name is Flatnose or "Romo," a word that indicates his face, particularly his nose, was damaged by syphilis. In contrast to the butcher, Berganza resolutely avoids the suggestion of a sexual encounter by saying he will not touch her fair hands. Berganza's final decision to leave the butcher master indicates that he refuses to participate in covetous activities, repudiating his role as Celestina figure in the sexual liaison.

In portraying an animal that avoids libido, Cervantes inverts the symbolic connection between lust and animality found in important source texts. In Ovid, Acteon lusts after Diana and he turns into a stag. Recalling the same episode in *The Golden Ass*, Lucius turns into an ass after he looks at a statue of Acteon staring hungrily at Diana with horns up (Graf 79). Berganza may be really a dog or a human that might have been metamorphosized, but no "animal" lust enters into his mind. Indeed, Cervantes' elimination of animal libido from the metamorphosis story reveals an aspect of the story that contributes to further nuance psychoanalytic interpretations of the dog, such as Ruth El Saffar's interpretation and even Freud's own blind spots when reading the story. It confirms that the oral telling of the dog dialogue itself, more than the turn to a subconscious animal per se, aids the ensign in the recovery of the past trauma of his deceitful marriage.[8] The ensign's redemption and recovery does not depend on making the animal a figure in the subconscious. In not connecting animality with sexual desire, Cervantes anticipates postures from Animal Studies that separate animality from the unconscious, such as Wolfe's reassessment of the animal in Freud or Derrida's reassessment of animality in Lacan ("And Say").

In positioning the butcher as malicious, Cervantes also portrays Berganza as exemplary in the sense that he rejects and abandons his participation in an animal spectacle, that is, the nascent Spanish bullfight. In their colloquy, Berganza tells Cipión that baiting bulls was the first thing he learned to do as a puppy. He explains:

---

[8] Ruth El Saffar argues that the ensign's written story (the dog dialogue) and its subsequent reading to his friend enable the recuperation from his past marriage based on lies, reflecting the therapeutic aspect of writing as a redemptive act that liberates one from trauma.

> My first master went by the name of Nicky Flatnose—a strong hothead, like all butchers. This Nick trained me and some other pups how to go with the older *alanos*, attack a bull, and make ourselves a prize of his ears. In no time, I became expert at this.[9] (27)

> El primero que conocí por amo fue uno llamado Nicolás el Romo, mozo robusto, doblado y colérico, como lo son todos aquellos que ejercitan la jifería. Este tal Nicolás me enseñaba a mí y a otros cachorros a que, en compañía de alanos viejos, arremetiésemos a los toros y les hiciésemos presa de las orejas. Con mucha facilidad salí un águila en esto. (302)

Berganza's description of the butcher and his shop portrays the atmosphere of the proto-bullfight. The slaughterhouse is filled with people before sunrise: "Every morning on meat days, all these toughs and their molls show up before dawn" ("Todas las mañanas que son días de carne, antes que amanezca, están en el matadero gran cantidad de mujercillas y muchachos" [27; 303]). The customers who arrived early would not only have gotten choice cuts, but also would have watched the slaughter spectacle that depended on the successful combined forces of the butcher and his dogs.

Because of their early role in bullbaiting and the bullfight, dogs constituted an important component in the genealogy of the definition of the human in the Spanish context from the sixteenth through the nineteenth century. Early texts indicate that the *alano* was bred as a holding animal for large game. Sebastián de Covarrubias writes under the entry *alano* that this dog is bred to take down big

9 I have altered the original by replacing "mastiff" with *alano* (the English breed of the mastiff has been used in translations of "The Dialogue of the Dogs" since the seventeenth century). Altering the translation from "mastiff" to *alano* is more than merely a simple lexical correction because the English openly changed the large Spanish war dog into the distinctly English breed known as the mastiff. The linguistic shift from *alano* to mastiff forms part of the English piracy project of the Spanish ideological construction of empire (that accompanied the creation of the English spaniel). Samuel Purchas' translation of the Spanish New World accounts in twelve volumes (1635) contains a description of Vasco Nuñez de Balboa's Bercerrillo. In his translation of Oviedo, Purchas writes: "Our mastiff dog Bercillio (sic) overtook one of them ... and pulled him down, having bit him cruelly..." (qtd. in Mastromarino 17). England has moved in the place of Spain as imperial power and, even though its Spanish name ("Becillo") has been mis-transcribed, the breed of the conquest is an *English* dog, the mastiff. Oviedo had exploited Charles V's motto, *plus ultra*, in that, as historian of the New World (his self fashioning as a new Pliny), he equated his work with the imperial expansion of Spain. A century and a half later, the Royal Society in England usurps the Spanish interest in animals and history in the New World as Joseph Glanvill would re-appropriate Charles V's motto in a new Anglo-scientific context with the publication of *Plus Ultra or the Progress and Advancement of Knowledge since the Days of Aristotle* (Asúa and French 221–2). The linguistic inflection in the use of the spaniel and the mastiff often re-fashioned Spanish canine semantics within a distinctly English imperial formulation. Also see Beusterien, "Goya's." For further reading the English appropriation of the Spanish, see Fuchs.

game and the bull: "these dogs are taught to grab the bull or boar by the ear" ("… tiene enseñados a estos perros a que asgan el toro o el jabalí de la oreja" [my trans.; 64]). Sources generally connect the *alano* with the bull. One contemporary of Covarrubias writes that *alanos* "are not corpulent dogs, but robust and powerful. They have such strength that they will hold on the ear of the bull not letting it free until it gives up" ("perros no muy corpulentos, pero robustos y recios, de tanta fuerza, que asiendo de la oreja a un toro, no lo soltaban hasta rendirle" [my trans.; qtd. in Martin 1563]).

In the early modern period, the training of the *alano*, however, did not begin out in the forest with noblemen on the hunt, but in the slaughterhouse. The *alano* was designed to aid the butcher in the slaughter of his product. An early reference to *alanos* in slaughterhouses occurs in the early sixteenth century when Herrera writes about the pastoral Spanish mastiff, stating that, in contrast to his interest in describing the sheep dog, he is not at all interested in hunting dogs or "*alanos* that belong to butchers" ("alanos de carniceros") [my trans.; 350]). As opposed to Herrera's focus on the pastoral, the sixteenth-century Barahonda hunting dialogue takes a special interest in the *alano*, providing a detailed explanation of why it was bred in the slaughterhouse:

> *Alanos* should be raised in abattoirs, butcheries, or slaughterhouses and, if possible, be fed the blood of bulls and cows so that they have a desire and are greedy for meat and blood. After that, they should be placed with other *alanos* already trained in hunting big game. In this way, they will not become spooked when they are placed in situations that they are not used to, such as when the other dogs get excited or are in danger. *Alanos* should never have their own bed but be raised on a hard surface so that they are tough and so that they become fine work animals. (my trans.)
>
> [Los *alanos*] se han de criar en los rastros, carnicerías o mataderos, de suerte que cebados en la sangre de los toros y vacas se hagan golosos y codiciosos de carne y sangre, y de allí se han de sacar en compañía de otros *alanos* ya diestros en las cosas del monte, porque no teman de acometer puestos con los que les dan ánimo en los trances y peligros que ellos no se han visto, y no dales jamás cama regalada, sino que se crien en aspereza para que sean duros y hechas a trabajos. (466)

Since they were bred to be greedy and gluttonous for meat, this description provides further proof of Berganza's exemplarity in abstaining from eating the meat. Moreover, the description aids in providing a more accurate picture of the context of early bullfighting since, while trained for hunting big game, the designing of the dog for butchers developed into a spectacle of its own. Documents from Seville give early signs that people gathered to see how the dogs and their master brought down the bull. One sixteenth-century document describes how bulls were caped in the corrals of a slaughterhouse (Marvin 62).

Written at the beginning of the seventeenth century when the bull spectacle increasingly became a more popular pastime, the anonymous poem "El Alano"

(1607) mentions the dog's important role in pulling down the bull and the poet elegizes his dog named León.[10] The modern Spanish bullfight developed out of two bullbaiting traditions. One tradition was connected with the elite culture of aristocratic games based upon the use of the horse. The other was carried out by commoners who used the dog to aid them in dominating the bull (Shubert 7). The bullfight from its earliest moments in its popular manifestations (and not part of the aristocratic taurine celebration) has butchers as bullfighters accompanied by their trusty dogs. Many of the most famous and earliest *toreros* were employed or had been employed in a slaughterhouses (Marvin 62). As the spectacle became an organized event in the eighteenth century, dogs took on an established ritualized role in the eighteenth century by being set on cowardly bulls. As once source states: "If nothing can excite his courage... enormous large dogs are let loose on him, who get hold of his neck and ears" (qtd. in Shubert 136). In his 1816 etchings entitled "La Tauromaquia," Goya portrays dogs doing the job of modern day picadores and banderilleros, reducing the strength of the bull. In one Goya etching they take down a bull.[11]

[10] Compare the León from this poem with a dog named León from New Spain. Conde de Galvez, whose dogs were used to rile up the bulls during the *corrida*, mourned León when he was killed in a bullfight on January 24, 1788: "he was killed by one of the bulls, a heartfelt loss to his owner who valued him for his ferocity" ("fue muerto por uno de los toros, con gran sentimiento de su dueño, que mucho lo estimaba por su ferocidad" [my trans.; Rangel 192]).

[11] For the bullfight in Goya, see Alves 199n.51. Further study is needed on the history of the nineteenth-century prohibition of dogs in bullfights. Colonial enlightenment thinker José Joaquín Fernández de Lizardi wrote *The Conference of a Bull and a Horse* (*Los currutacos herrados y caballos habladores*) (1811) in which he critiques the irrationality of humans for killing bulls for entertainment when they would be more benefit in agriculture (Alves 198). In Spain, ever since the mid-nineteenth century, the use of dogs in bullfights was referred to increasingly to as "brutal custom" (qtd. in Shubert 137). Prohibitions of dogs in bullfights occur in the early 1880s (Shubert 137). Jacinto Benavente's "The New Dialogue of the Dogs" ("Nuevo coloquio de los perros") describes the decline of dogs in the bullring: "until a delicate feeling of humanity decided to substitute the dog for banderillas with fireworks attached, not necessarily giving an advantage to the physical well-being of the bull, but, nonetheless, an advantage for the bull's mental well-being" ("hasta que un delicado sentimiento de humanidad los sustituyera por las banderillas de fuego, sí no con gran ventaja para el toro en el físico, si para su conciencia" [my trans.; n.p.]). In contrast to the inhumanity of the use of the dog, twentieth-century writers understood the bullfight as a spectacle that represented humanity: "In the bullfight the Spaniard has found the most perfect expression for defining his human quality" (qtd. in Marvin vii). In the twentieth century it became a *fiesta nacional* (the first comprehensive national regulations were approved in 1917 and Franco's Spain would later make it a national pastime) in which the bullfight is associated with an authentic masculine and uniquely Spanish character.

One important characteristic of early modern anthropocentrism was the use of the large dog's body as an icon in early modern spectacles.[12] While in early modern England the mastiff baited bulls, the impact of the animal's role in spectacle would be especially important in Spain. Spain did not use the mastiff, a distinctly-designed English breed, but the *alano.* Since the second half of the sixteenth century, the spectacle of the large dog's body in the Spanish bullfight served to showcase a definition of the human. The human was both animal—unleashed fury, wild masculinity, and uncontrollable rage—and the being that dominated the animal by demonstrating exquisite control over the fury, by containing wild masculinity, and by ultimately destroying the uncontrollable rage. Although their role was not part of a scripted bullfight in the orchestrated twentieth-century ritual, dogs from the sixteenth through the nineteenth century teamed up with the bullfighter and bull killer from the first moments when the event became a popular spectacle. In the eyes of those who go to the spectacle, if the bull was the wild animal, then the butcher-bullfighter and his dogs were coded as the human that controlled the unleashed animal fury of the bull.

At the end of the episode, Berganza forgets the programmed training and human design that made him an accomplice to the butcher bull fighter. When he rejects his role as butcher's accomplice, Berganza refuses to participate in the bullfight, perhaps the most renowned Spanish anthropocentric tradition of animal design. Berganza no longer acts as an appendage to the human body in overcoming the forces of wild animal fury. He transforms himself in what he hopes will be a canine with the more peaceful profession of the *mastín* or pastoral sheep dog, a breed that—as Herrera points out—had a radically different purpose than the *alano.*[13]

The final way in which Cervantes portrays Berganza as exemplary in the first episode is by making him critique the butcher for the crime of being a killer. Ostensibly, Berganza voices the classic critique of the murderer who makes no distinction between the animal he kills and the person he murders.[14] Berganza

---

12 Cervantes treats the popular version of the bullfight as opposed to the aristocratic taurine spectacle. Aside from the aristocratic version of the bull spectacle, Cervantes also does not mention the aristocratic simulacrum of the hunt, another important dog spectacle that often used *alanos.* More research is needed on the enclosed simulacrum of hunting spectacles for dogs for public and noble amusement. For a visual example of the simulacrum of the aristocratic hunt with dogs, see Mateos, Lámina 5, 87 and Velazquez's *Philip IV Hunting Wild Boar*. For a description of a simulacrum spectacle of the hunt in New Spain, see Rangel 29.

13 Herrera's implicit critique of the hunt is explicit in many early modern Spanish texts (Greer, "Diana").

14 Plutarch presents a similar conflation on the ease with which butchers will kill both cows and men when arguing for vegetarianism (Thomas 328-33). Other ancient sources argue on behalf of abstinence of flesh-eating and vegetarianism, including Seneca, Porphyry, Petrarch and Ovid, a source that Cervantes would have surely known (Newmyer). Thomas cites the connection between early English critiques of butchers and early modern vegetarianism (294). I have found very little on case histories that critique butchers in the early modern Spanish context. Note, however, a legal case in Spain that cites a butcher as murderer (Mckendrick 314).

states: "But nothing amazed or disgusted me more than how these butchers kill a man as easily as you would a cow. In an instant, two or three of them plunge their horn-handled dirks into someone's belly as if they were goring a steer" ("Pero ninguna cosa me admiraba más ni me parecía peor que el ver que estos jiferos con la misma facilidad matan a un hombre que a una vaca; por quítame allá esa paja, a dos por tres, meten un cuchillo de cachas amarillas por la barriga de una persona, como si acocotasen un toro" [28, 303]). But with this episode Cervantes forces the reader to do more than read the episode as simply a tale that condemns human murder. The condemnation of the butcher as a "murderer" does not make sense in the context of the animal exemplum because that condemnation comes from a dog or a human in dog's form. Cervantes forces a simple question that unsettles the definition of murder based on the human-animal divide: since Berganza apparently is an animal, would killing Berganza also be equivalent to murder or would it be killing an animal with a differently-based ethical protocol?

Berganza's butcher critique impels the reader to conceive of the animal as having unique demands, moving toward an ethical imperative that nurtures living. Donna Haraway has studied the dog and shared animal-human suffering in order to destabilize human exceptionalism and to evaluate what is at stake when the animal is agent and actor (*Companion; When*). Donna Haraway writes in *When Species Meet* that animals like the dog are one of the many creatures that co-shape and co-create the human. Because of the interdependency of the animal and the human, Haraway insists that humans should come to share the pain of the animal in order to act ethically. For Donna Haraway, every atrocity merits its own potent languages and ethical responses: "we need other nouns and pronouns for the kin genres of companion species… ."[15] She points out that one of these nouns is "killing" since animals are killed while humans are murdered. Thinking through "killing" can help achieve responsibility: "if all were not made 'killable' then we could learn to live responsibly within the necessity and labor of killing.…" (81). After giving the example of David Lurie in *Disgrace* who takes stray animals to their death, Haraway writes that "I do not think we can nurture living until we get better at facing killing" (81). Cervantes joins Haraway in forcing a thinking-through of the meaning of killing by having Berganza, a creature intentionally defined as a blur between the animal and human, criticize a man who does not distinguish between the animals and people that he kills.

While Cervantes uses the figure of Berganza to muddle the classic humanist critique of the murderer and killer, he clearly follows the tradition of animal exemplarity with Berganza's behavior in the first episode. The butcher is evil and Berganza is good because he avoids greed, avarice, lust, participating in the bull spectacle, and killing. Indeed at the conclusion of the episode Cervantes explicitly connects Berganza with a saint by having him go through the Seville neighborhood of Saint Bernard when he abandons the butcher. Medieval exempla generally re-configured Aristotle's notion of mimesis into a Christian context. The

[15] Haraway also sees the Animal Studies project as connected to her feminist one. The complete quote reads: "we need other nouns and pronouns for the kin genres of companion species... just as we did (and still do) for the spectrum of genders" (95–6).

exemplary animal, therefore, did not just provide a model for the good, but led one to redemption and salvation like a saint. The suggested connection between saints and dogs (both Bernard and Domingo) is even further heightened since in the medieval tradition Bernard's mother was thought to have given birth to a dog.

After leaving the butcher, Berganza describes another adventure in which he portrays himself as exemplary. As in the first episode Berganza refrains from eating the malicious butcher's meat, so in the second episode he is a model of pacifism, refraining from and condemning the criminal slaughter of sheep. In this episode, Cervantes distinguishes Berganza from his shepherd masters, who are wolves among men. The dishonest shepherds tell Berganza that wolves are coming and slaughtering the sheep at night. Berganza, however, discovers that the shepherds themselves are the "wolves" since they are massacring the sheep. Berganza explains: "I gasped when I saw that the shepherds were the wolves, and were raiding the flock they were supposed to guard" ("Pasméme, quedé suspenso cuando vi que los pastores eran los lobos y que despedazaban el ganado los mismos que le habían de guardar" [37; 311]). As in the case of his experience with the butcher, Berganza decides he must abandon the immoral band of shepherds.

Beyond making Berganza exemplary, this episode, like the first one with the butcher, confuses the logic of the animal-human divide. Berganza condemns the shepherds for being wolves. But who are the real animals here, Berganza or the shepherds? By having the unclearly defined Berganza criticize humans as wolves, Cervantes again questions another classic humanist position—that humans live like wolves. Cervantes' reference to and disordering of the logic of the *homo homini lupus* tradition in the second episode is a profound reaction to a classical conception of humans and animals that was revived in the early modern period. The trope of *homo hominis lupus* is found in the opening pages of *La Celestina.* The author of *La Celestina* suggested that all live in a dog-eat-dog world when he writes that nature is continually at war with itself. The author describes a series of animal species at war for two pages: "Among animals the same constant struggles are to be perceived. Fish and fowl, wild beasts and serpent—the species all pursue each other: the lion—the wolf; the wolf—the goat; and the dog—the hare" ("Pues entre los animales ningún género carece de guerra: pesces, fieras, aves, serpientes, del cual todo, una especie a otra persigue: el león al lobo, el lobo la cabra, el perro la liebre" [9, 198]). The author of *La Celestina* suggests the same conflict is present in human wars, enmities, envies, compulsions, movements and dissatisfactions (11).

In the seventeenth century, Baltasar Gracián would influentially perpetuate the same vein of thinking in political philosophy by paying special attention to the classic adage *homo hominis lupus* in *El criticón.*[16] Political theorists, especially Hobbes, asserted that an animal or natural state—*homo hominis lupus*—reigned as an epoch prior to the arrival of a sovereign and the foundation of the city. Like Freud who assumed that the human appeared *after* the animal, so Hobbes' theory supposes a mythic and spurious moment when the sovereign surfaces *after* a period in which man is a wolf to men and nature is the war of all against all.

---

[16] Abel Alves notes anthropocentric instances in Gracián in which he uses the animal as figure; he creates animal metaphors for human behavior (55).

In the same way that the Freudian origin myth is ungrounded through faulty temporal logic, so Animal Studies scholars have dispelled the supposed temporal logic of an assumed political state and sovereign as needed *a posteriori* to a strife-ridden animalized natural state. By having the undefined human-animal Berganza take the judgmental position that shepherds are living like wolves, Cervantes' evokes *homo homini lupus* and questions the role of the animal as figure for the human in the adage. When a human being, like the author Gracián, condemns men for living in an animal state, he implicitly positions himself as non-animal and rational. But when the fictional Berganza condemns men for living in an animal state, he may be positioning himself as rational, but the ontological status of the human is called into question. In this sense, Cervantes not only re-conceptualizes the role of the wolf metaphor in the classical conception of a society characterized by animal-like strife, but he marks an important step toward the manifestation of the Animal Studies arguments against the use of the *homo homini lupus* metaphor that includes not just a critique of Hobbesian political theory, but also Freudian and Lacanian psychoanalytic theory.[17]

Despite the exemplary clarity of Berganza's behavior in contrast to his *amos* in the first two episodes, Cervantes intentionally tangles the association between animals and positive and negative exemplarity from the outset. On one hand, Cervantes writes that Cipión and Berganza are "two guard dogs who go around at night carrying lanterns for the monks, to give them light while they collect alms" ("dos perros que con dos lanternas andan de noche con los hermanos de la Capacha, alumbrándoles cuando piden limosna" [17; 293]). With this reference, Cervantes evokes the commonplace notion from the day that the dogs are connected to the pursuit of an exemplary life associated with Saint Dominic and his order. A multitude of images from the sixteenth and seventeenth century can be found of Dominican dogs holding torches in their mouths such as Claudio Coello's *Santo Domingo de Guzmán* which hangs in El Prado.[18] The lexicographer Francisco del Rosal makes important observations on dogs and Dominicans with respect to their exemplarity for humans. Rosal's description of the Dominicans states: "the preaching order of the glorious Saint Dominic has the dog with a torch in its mouth as its coat of arms, a symbol of preaching and representative of the purest doctrine and an exemplary life" ("la orden de predicadores del glorioso Santo Domingo

---

[17] Derrida, Agamben, and Wolfe link the political critique of animality in the foundation myth of a natural world in strife to a critique found psychoanalytic theory, especially following *homo homini lupus* in Freud's *Civilization and Its Discontents.* Derrida, citing Gracián, explains that an untheorized animality exists in Lacan by stating that the adage *homo homini lupus* means that just because animality or instinct is contained *in* humans does not mean the animality is more contained ("And Say").

[18] The sixteenth-century Dominican iconography—a dog holding a torch in its mouth—is not only found in paintings, but would form part of the architecture in New World churches. See for instance the convent of Saint James the Apostle at Cuilapan in Oaxaca, Mexico that has two dogs with torches in the mouths flanking the Dominican coat of arms.

tiene por blasón perros, símbolo de la predicación, con antorchas en las bocas insignias de la doctrina purísima y vida ejemplar" [my trans.; 60]).[19]

On the other hand, completely contradicting the dog's connection to saintliness, Cervantes simultaneously links the dogs to a tradition in which they are icons of the diabolic. The dog was not just a medieval icon for sins like avarice and gluttony, but it was equated with Satan. Cervantes intentionally never answers a fundamental question in the dialogue: the dogs' mother may be a witch. As opposed to his connection with Saint Bernard and Saint Dominic, Berganza also has possible dark origins. In popular legend, upon initiation, witches were often given puppies by the devil, who himself might be a dog. Witches also sometimes had to change themselves into dogs to properly execute spells. The witch–dog connection in "The Dialogue of the Dogs" also evokes the tradition of Hecate, goddess of the night and necromancy, whose chief animal companion was a large black bitch that in many legends had the previous shape of a human. Goya sketches one witch transforming into a dog (or vice versa?) in *The Witch's Meal* (*Comida de las Brujas*). Literary and historic sources from Cervantes' period claim that the devil used the dog's body for its evil purposes. One episode of two black dog demons can be found in Antonio de Torquemada's *The Spanish Mandevile of Miracles; Or The Garden of Curious Flowers* (*El jardín de las flores curiosas*). Inquisition records also report that Basque witches changed into dogs in 1609 and 1614 (Alves 114).

Aside from their suggested infernal origins, the existence of speaking dogs was a direct sign of the presence of the devil. A common belief from the day supposed that talking dogs were instruments through which satanic forces spoke.[20] Luis Pérez's *Concerning the Dog and the Horse* (*Del can y el caballo*) (1568)

---

[19] The word *can* in *dominicano* (Dominican) is not connected etymologically to dog. *Dominicano* literally is from the Latin word for God or Lord. Nonetheless, "Domini*cano*" had a firmly connected folk connection to the Latin *can* (dog). The Dominicans were associated with the Inquisition and considered the "Dogs of God," hounds in pursuit of heretics (Cárdenas). Ironically, the objects of their pursuits were often referred to as "dogs" even as some Dominicans such as Martín de Porres made it part of their mission to care for and comfort *real* dogs (Alves). Many other stories connect dogs with saintly exemplarity. Like the case of Saint Dominic, popular saint legends reputed Saint Bernard's mother to have given birth to a dog. The story of Saint Guinefort is a fascinating instance of an animal exemplum in which the dog itself was venerated as a martyr and healer, and was canonized (Schmitt).

[20] A variety of sources associate dogs with negative exemplarity by connecting them to the devil. The devil that used the dog as vehicle for its voice can be found in the medieval tradition, such as in Vitoria's *Concerning Magic* (*De la magia*) and the works of Saint Isidore that were re-reprinted in early modern Spain (Alves 126). A contemporary of Cervantes, Benito Carrasco, wrote a short poem entitled "The Devil is the Face of a Dog." In New Spain, one engraving of Saint Rose, the first American saint, shows a devil incarnated into a dog biting her midsection (Pérez, "Dogs"). For more on animals connected to the Satanic, see Alves. One early modern English poem puts the dog in cahoots with the devil and witches. In "A Dog's Elegy or Rupert's Tears" (1644), the Pope, the devil and a witch are called the "close-mourners" at the dog's funeral (Booth).

reports that it is certain that "a dog cannot talk" ("el can no puede hablar") and, if it does, it is "the tricks and deception of the devil" ("fraude y engaño del diablo") (my trans; 21). *The Garden of Curious Flowers* records that demons and spirits entered all sorts of creatures such as monkeys or dogs and, upon entering the creature, they use the animal "as their own instrument through whose tongue they speak and say what they want" ("como instrumento propio, y por su lengua hablan y dicen lo que quieren") (my trans.; 303–304).[21] The creation of the ambiguous dogs destabilizes the Aristotelian mimetic precept since the reader cannot tell if they represent virtue or vice. Cervantes playfully suggests that they are exemplary like the Dominicans, but also are instruments of the devil. They *may* be saintly or they *may* be instruments through which the devil deceives and leads astray.

While some critics may assume that the butcher and shepherds are "inhuman" or "inhumane," Cervantes does not use either word to modify them. Even though he borrows the ethical exemplarity of the dog from many sources for the first episodes, Cervantes refrains from using the word "inhuman" to describe the masters even though it can be found in source material for the tale. In the "Dialogue between Two Little Dogs" ("Diálogo entre dos perrillos") (ca. 1585) by Baltasar de Alcázar, the dog Zarpilla, not Berganza, provides a picaresque autobiography. This source text directly makes out the human master to be *inhuman* in contrast with the exemplary talking dog. Zarpilla, the canine protagonist, begins the poem claiming: "I am a little *gozque* from Seville / owned by an inhuman mayor / and I wish I weren't his" ("Gozquijo soy sevillano / y de un alcalde inhumano, / que ojalá no fuera suyo" [my trans.; 426]).

In a much more complex maneuver that questions the human, Cervantes avoids the denomination of "inhuman" for the butcher and thieving shepherds. He jumbles up the relationship between the animal and the human and the tradition of the animal exempla by intentionally suggesting the dogs are simultaneously iconic sinners and saints. "The Dialogue of the Dogs" is the last novella in Cervantes' collection entitled *Exemplary Novels* (*Novelas ejemplares*) and although Berganza is an exemplary figure in the first two episodes, the dialogue is not a conventional medieval exemplum in which the animal is model figure through which the human can learn to do good. Indeed, after Berganza concludes the tale of the deceitful shepherds, Cipión reacts to his moralizing and tells Berganza: "But stay with this, since I don't want us to seem like some know-it-all preachers." Following Cipión's remonstration, the exemplarity of the dialogue takes a turn and Berganza's apparently clearly-established fixed exemplarity—in which the animal operates as model for good behavior—falls apart. Indeed, following the first two episodes wherein Berganza operates as exemplary, Cervantes then inserts the contradictory tradition of animal exemplarity into the same dog story and same animal character. In this way, "The Dialogue of the Dogs" turns to the animal to

[21] Cervantes knew Antonio de Torquemada's *The Garden of Curious Flowers* (1570); in *Don Quijote,* it is one of the books taken from Quijote's library and thrown in the fire.

crumble the scaffolding that supported the clearly delineated purpose of the two types of exemplary tales.

If Cervantes had ended the dialogue after the second episode, his dialogue would have fit much more neatly in the traditional animal exemplum in which the dog provides a model toward which the reader should strive to act ethically, not like an animal, but a saintly human. But in contrast to the tradition in which the animal shows the human his evil ways, Cervantes switches the terms of exemplarity in the episodes that follow by portraying Berganza as a negative model. In the episode with his third master, the merchant, Berganza, judges an Afro-Hispanic woman's dark skin to be a sign of sin and lust and he also violently and brutally attacks her. Berganza describes the incident: "One inky night, as she was coming down for her usual pastime, I went after her without barking, so as not to disturb the household, and in an instant I tore her shift to shreds and gouged her thigh" ("bajando una noche muy escura a su acustrumbrado pasatiempo, arremetí sin ladrar, por que no se alboratasen la casa, y en un instante le hice pedazos toda la camisa y le arranqué un pedazo de muslo" [54; 322]). Berganza continues: "That little manoeuvre sufficed to keep her bedridden for more than a week, faking I don't know what illness for her masters. She healed up, returned another night, and I tangled with the bitch again. Without biting her, I clawed her all over as if I were carding a fleece" ("burla que fue bastante a tenerla de versa más de ocho días en la cama, fingiendo para con sus amos no sé que enfermedad. Sanó, volvió otra noche, y yo volví a la pelea con mi perra, y sin morderla le arañé todo el cuerpo como si la hubiera cardado como manta" [54; 322]). Cervantes directs the reader's sympathies toward Berganza until the violent attack upon this woman. Berganza also suggests that she has the skin of a sheep and calls her a bitch. By highlighting his violence and arrogance, Cervantes suggests that Berganza may not be a saintly model, but a negative one. Indeed, Berganza's animalization of the woman makes the reader question the moral logic behind expressions that animalize humans and, indeed, the role that the animal has in how humans define themselves as human.

Berganza again treats a woman in animal terms when he says that the witch Cañizares has an ugly animal belly and breasts: "Her belly, which looked like cowhide, covered her privates, even hung halfway down her shanks. Her breasts looked like two dried and wrinkled udders" ("la barriga, que era de badana, se cubría las partes deshonestas, y aun le colgaba hasta la mitad de los muslos; las tetas semejaban dos vejigas de vaca secas y arrugadas" [83–84; 344]). The witch, in Tobias Foster Gittes' terms, exhibits "the noblest aspects of human behavior" (356), but Berganza treats her violently, dragging her naked out of her cell into public view and by subjecting her body to public mockery and prodding.[22] Cervantes once again portrays Berganza in negative terms, forcing one to question animal exemplarity. Moreover, Berganza once again animalizes a woman. The

[22] Foster Gittes rightly points out the negative exemplarity of Berganza in the scene, but in his study of the dialogue he reverts to the traditional logic of basing exemplarity on "human behavior."

dialogue, therefore, does not logically allow the terms of traditional animal exempla based on an animal-human hierarchy to stand.

Just after Berganza behaves as the model human in the butcher and shepherd episodes, he turns and uses human-animal hierarchical logic to inflict violence on the black servant and the witch. Cervantes intentionally undermines the purpose of the two classic animal exempla by giving a contradictory valence to the character of Berganza. In the very last episode when Berganza is collecting alms with Mahudes, Cervantes further defuses the function of mimesis in the traditional animal exemplum by creating a new exemplary genre. In this episode, Cipión learns from the animal exemplum as told by Berganza. Typically, in the animal exemplum humans imitate proper ways of acting or learn what to avoid through the figure of the animal. We know how to act or not act because we can either act *like* or *not like* the animal. But at the very conclusion of "The Dialogue of the Dogs," Berganza's animal story itself is one in which the animal is compared to the human world rather than one in which the human is compared to the animal. Berganza describes a smug little lap dog in the arms of a lady: "The creature saw me and jumped from its mistress's arms, chasing me with a great barking fury, and didn't stop till it bit me in the leg" ("la cual, cuando me vio, saltó de los brazos de su señora y arremetió a mí ladrando, y con tan gran denuedo, que no paró hasta morderme de una pierna" [105; 358]). Cipión responds by stating that the little dog is *like* those servants who have a haughty attitude because they are under their master's protection: "You can see it in those little men who let fly with their insolence—all the while basking in the shade of their masters" ("Una muestra y señal desa verdad que dices nos algunos hombrecillos que a la sombra de sus amos se atreven a ser insolentes" [105; 358–9]). The animal exemplum disappears. Cervantes creates a "human" exemplum for animals since Cipión, not a human, learns from the episode. Moreover, he uses a simile taken from the human world—the lapdog is like "those little men"—to arrive at the moral of the final episode for his own canine, not human, behavior.[23]

With the dog dialogue, Cervantes writes a new animal exemplum. He destabilizes the terms of human and animal exceptionalism that formed the base of the two generic forms of animal exemplum literature and renders the logic of both as inconsistent and inoperative, abandoning the terms of the ethics of the animal-human divide. If one reads the first episode of "The Dialogue of the Dogs" as a contained story, one might read it in the context of the Aristotelian precept that one can learn from the exemplum to avoid vice since the butcher fits the classic portrayal of the sinful retrograde. But, taking the entire dialogue into account, Cervantes urges the reader to ask what happens to mimesis and ethical behavior when they are not based on the terms of the animal-human divide.

---

[23] As an aside, one recent scientific study that appeared in the press concluded that dogs can be ethical based upon how they watch other dogs. The study showed that when one dog is given a treat for a trick, but another is not, then the first dog will become upset that the second dog is not rewarded (Wolchover).

Critics of the dog dialogue have read it as revealing biographical information about Cervantes. Some argue that it reveals Cervantes to be a man who was intolerant toward others while, paradoxically, others argue that it shows that he was tolerant toward human others, especially the Moors.[24] Both of these readings do not understand the animal as unique other. In articulating the animal's role as "other" other, "The Dialogue of the Dogs" should be read as disposing of certain foundational anthropocentric precepts that connected animality with libido, languagelessness, and an uncivilized state of nature. In contrast to an Aristotelian mimesis based on an animal–human divide, Cervantes provides important bases that will be further developed in Animal Studies criticism, especially by Jacques Derrida, Cary Wolfe, and Donna Haraway. In this way, critics of the dialogue can reach outside of an anthropocentric critical cycle and can show how "The Dialogue of the Dogs" does not rely on the maintenance of the animal–human divide in ethical decision-making. It rescinds the semantics of the animal-versus-human to arrive at its exemplum purpose and, what I will suggest in the next chapter, a more fully operative notion of justice.

[24] Amezúa interprets Cervantes as anti-Moorish. In his meticulous philological work on "The Dialogue of the Dogs," Amezúa ties his "humanist" method to an idea of the human as linked the ideological apparatus of the *falangistas*, the nationalist, Catholic values of Franco. In his 1965 opus, Amezúa calls Cervantes a "pureblooded, old stock gentleman and a passionate Spaniard" and "a good Spaniard who hates dirty Moors" ("viejo hidalgo de limpia sangre y español apasionado" y un "buen español que abomina de la morisca canalla" [my trans.; *Cervantes* 439]). In radical contrast, Alban K. Forcione argues that Cervantes should be considered a man of humanist tolerance. Forcione's concludes his discussion of "Dialogue of the Dogs" by arguing that Cervantes real affinities lie "with the tolerant, pluralistic vision of human possibilities and limitations that was characteristic of Erasmist Christian humanism" (*Cervantes and the Mystery* 214). Some critics have recently followed Forcione in asserting that Cervantes held a pluralistic vision toward the Moors in general. William Childers argues that Cervantes gives narrative form to the *morisco's* ambivalent identities (177) and Eric Graf argues that Cervantes offers a manifesto on behalf of the oppressed *moriscos* (22).

Chapter 3
When the Dog is a Book: A Post-Human Ethics in Cervantes

> "I ask myself what is the most immediate sense of the word ethics, in what way is it already other than morality, well, ethics is better known to us today under another name, the word ethology."
>
> —Gilles Deleuze

In many ways, the creation of the shape-shifting Berganza is Cervantes' way of abandoning the design of the *alano*, a breed that had quickly achieved the status of the Spaniard's "best friend" in the sense it was a projection of growing masculine and national fantasies, particularly as co-protagonist with butchers in bullfighting, but also with nobleman hunters and conquistadors. Cervantes designs a dog that ultimately abandons the logic of the ideal exemplary big dog that occupies the inside ("us", the Spanish, the good, the masculine). But he also does not design the alternative dog model, that is, the brutish, diabolical little dog that occupies the outside (them, the non-Spanish, the bad, the feminine). In the end, Berganza is neither large nor small—neither virtuous nor vicious—because Cervantes blurs the ethical imperative of the classic exemplary tale and, in turn, blurs the logic of the friend and enemy because the dogs simultaneously represent saintly big dog ideal behavior and, on the other hand, demonic small dog behavior. Berganza is ultimately an ambiguously-defined creature "in between" and "in the middle." The termination of the large-small dog dichotomy in "The Dialogue of the Dogs" puts Cervantes on an evolutionary trajectory toward the birth of a new dog in literature, the portrayal of the medium-sized dog also known as the *podenco*.

Philosopher of ethical action Immanuel Levinas writes that the ethical imperative towards the other is based on a theory of human, not animal, friendship.[1] In Levinas, we act toward the other because of its similarity to humans—the other has a human face—and because of its impossible-to-define difference—the human face has something impenetrably unique beyond definition.[2] Love or sympathy

---

1 With respect to the animal in Levinas, H. Peter Steeves writes: "he carefully pushes away before the question can truly be served" (22). Also see Steeves for more information on Levinas' shortcomings with respect to thinking about the dog.

2 Early modern Europe provides some important historical examples that elucidate how the human face uniquely connects to reason and heightened language, providing a partial genealogy to the notion of the face in Levinas. These include: 1. The revival of the Aristotelian idea that only the human, not animal, face reflects the capacity to reason. One poem from the sixteenth-century poet López Sedan interprets the Acteon myth based on the understanding of the human face. The face reflects the inclination of the soul and, since only the human face looks toward the heavens, it confirms man as the most sacred and noble of

for one's neighbor, fellow, or friend ("brotherly love") depends on an other that, although always different, is like a human, never like the animal. When Levinas extends his idea toward the animal, he does so under the rubric of the human:

It is clear that, without considering animals as human beings, the ethical extends to all living beings. We do not want to make an animal suffer needlessly and so on. But the prototype of this is human ethics. Vegetarianism, for example, arises from the transference to animals of the idea of suffering. The animal suffers. It is because we, as human, know what suffering is that we can have this obligation … with the appearance of the human—and this is my entire philosophy—there is something more important than my life, and that is the life of the other. (50)

Levinas' idea extends the idea of pluralism and ethical attention to include the animal, but, even though the "life of the other" includes animals, Levinas' empathy toward the animal exists because the human knows what *human* suffering is.

Cary Wolfe explains that the Levinas idea about pluralization, that is, ethical attention to human and other beings, fails to the degree that it does not destabilize or throw into question "the schema of the human who undertakes such pluralization" ("Human" 568). Abandoning the paradigm in which the human face centers ethical action, Wolfe and other Animal Studies scholars insist on conceiving the animal as a being with unique demands. If Levinas theorized the human through a notion of the other, then the animal must be theorized through the non-friend, a different other, or an "other" other. The "other" other is, at its root, non-oppositional and infinitely differentiated—not reduced to the general human subject versus the non subject that is animal. Under this model, the notions of "animality" and its iterations such as bestial and sub-human along with its opposite iterations such as "humane" are ineffective descriptive words in determining ethical action. For Wolfe, then, with the elimination of the animal and the human in the object-subject identification process, an ethics can take more fully into account the whole differentiated field of experience of a world of life-forms.

---

the creatures. In contrast to the human face that looks to the divine, when Acteon is a stag, he looks down: "When reason reigned / His face was held high, / But, when he lost reason, / His face lowered toward the earth" (Cuando la razón regía, / El rostro alzado tenía, / Mas luego que se perdió, / El rostro a tierra bajó ...) (my trans.; qtd. in Barahonda 87) In contrast with the animal and base humanity expressed in physiognomy tracts such as Rocca's, the Italian Neoplatonic philosophy Ficino recalls how the soul is like a mirror image of the divine face, an oft-repeated concept such as in Milton's phrase "the human face divine."

2. A connection existed between the human face and the ability to create poetic language. In his reading of the Acteon myth, Baltasar Gracián writes of how the experience of the human face is the basis of rich poetic expression. In referring to the Acteon myth, Gracián uses the notion of *carear*, a word that meant for him having been in the wrong place at the wrong time, or, more literally, a chance face-to-face encounter in which someone sees what she or he should not have seen, particularly *la cara* (face). Gracián writes that as Acteon saw what he shouldn't have seen, a human face (Diana's), so Ovid himself should not have seen a human face (someone doing something wrong at the Augustan court). Gracián writes that, because of his experience of *carear* and subsequent exile, Ovid creates a superb poetic metaphor (like Acteon torn apart by his dogs, so Ovid describes his entrails as ripped apart to describe the anguish of exile) (Schwartz Lerner).

As evidenced by "The Dialogue of the Dogs," Miguel de Cervantes was especially interested in the representation of animals, especially dogs. During his experience as a captive in Argel, his captors, in all likelihood, often called him a dog (Garcés 141). Dogs appear in his work, not just as protagonists of a dialogue, but in many other ways. In *Don Quijote,* the tongue-in-cheek Cervantes writes that the true author of the book is a *galgo*: "If any objection can be raised regarding the truth of this … and if something of value is missing from it, in my opinion the fault lies with the *galgo* who was its author rather than with any defect in its subject" ("Si a esta se le puede poner alguna objeción cerca de su verdad … y si algo bueno en ella faltare, para mí tengo que fue por culpa del galgo de su autor, antes que por falta de sujeto" [68–9; 110]).[3] The conclusion of Part II of *Don Quijote* mentions that when Quijote reaches his village and arrives home for good (to die), some *galgos* (greyhounds) chase a rabbit right between the feet of Sancho's donkey.[4] The opening line of *Don Quijote* also reminds the reader of one of the protagonist's former companions, an old *galgo corredor*. After Quijote leaves on his first sally, he never mentions the dog again.[5]

Aside from his general interest in dogs, Cervantes was interested in and influenced by renaissance modes of thinking and creation. "The Dialogue of the Dogs" takes as its point of departure the dialogues of the renaissance humanists (Hart). Cervantes, nonetheless, does not only radically reconfigure the animal exemplum and the humanist dialogue, but the renaissance representation of the canine. Like "The Dialogue of the Dogs," the "Prologue to the Reader" for Part II of *Don Quijote* provides a pithy and illuminative way to ascertain how Cervantes borrows renaissance conceptions of the animal only to reconfigure them. In this prologue, Cervantes justifies the writing of his book through two odd anecdotes. With the dog anecdotes in the "Prologue to the Reader," Cervantes makes reference to and rejects three anthropocentric representations of the dog: as part of a large-small discursive tradition; as object of carnivalesque ridicule; and as object toward which one should act humanely. In contrast to the beginning of *La Celestina* where the dog is a figure for the humanist enterprise of imagining the idea for writing the book, the prologue to Part II of *Don Quijote* literally makes the dog the book. In the creation of the dog-book, Cervantes forges a notion of ethics that does not rely on the human-versus-animal and that demands a different sort of empathy given toward the human friend or animal as understood through human ethical action. In

[3] All *Don Quijote* quotes in English are taken from the Grossman edition and Spanish quotes are taken from the Rico edition.

[4] More study is needed on the fascinating portrayal of Sancho and his donkey in Part II of *Don Quijote.* See Beusterien and Callicott.

[5] Ben Johnson provides one humorous example to contrast Don Quijote who forgets about his dog. In *Every Man out of his Humour* (1599), Ben Johnson buffoons Puntarudo, the "poore knight errant," who treats his hunting hound as his lap dog (Boehrer, *Shakespeare* 151–2).

short, the dog-book demands an empathy that is constituted through the inclusion of non-human faces that demand ethical responsibility.

Borrowing from the classical and medieval tradition, the early modern period in Europe bred and designed small and large dogs with a gendered and ennobling symbolics.[6] Europe represented the Spanish dog, the "spaniel," as small, while, in contrast, Spain represented the American dog as small. The Spanish invested the big dog with knightly, masculine values and associated the small dog with women and outsiders, most especially native Americans. Spaniards arrived to America with their own large purebred soldier dogs and considered American dogs as *gozques*, a term that implied both of a smaller size and impurity of *casta*.[7] The first description of dogs native to America from a European perspective occurred when, in a letter to Diego Alvárez Chanca from his second voyage (1494), Columbus called them *gozques* (Colón). The chronicler Oviedo also consistently called American dogs *gozques* (Alvárez López 413).[8] Inca Garcilaso de la Vega states that there are no

---

6 Classic and medieval texts relate nobility and gender to canine size. Large dogs in ancient times served the male in hunting and herding. Already in Homer's text from the ninth or eighth century before the common era, dogs were long performing their traditional roles as hunter (Argos was Odysseus' trusted hunting hound) and as guardians of livestock. In the fifth-century polis with its more advanced social groupings, dogs' lives changed in the sense they were not necessary for human survival and they did not necessarily have to give anything back to the human except companionship. Cuddly, curly tailed Maltese dogs occupied women's quarters, the place where the Athenian male also lived from birth to somewhere into early adolescence (Kitchell). Later in Rome, fondness for dogs was part and parcel of urban home life, particularly that of women. In Propertius one husband must yield his place on the marital bed for his wife's little dog and Juvenal wrote that Roman women would sooner let their husbands die than their dogs. In the medieval period large dogs are associated with nobility. A popular medieval sermon was translated in Spanish in the fifteenth century that describes dogs as either aristocratic or commoners. The author writes: "It is said that there are two types of dogs, some noble and others common. The noble ones are simple, quiet, and love solitude while the others are rude, base, bark all the time and wantonly wander about town squares" ("Dícese que son dos maneras de canes, unos nobles e otros villanos, y los nobles son simples e callados e amadoras de soledad e las que son aldeanas e villanas son vocingleras e andariegas e troteras de la plazas" (my trans.; *Espéculo* 98).

7 Attached to its lack of use value it was often the word used to denominate a dog with no owner or stray. For instance, in the novel *Guzmán de Alfarache*, *gozques* were synonymous with strays. The novel describes a hoard of yapping *gozques*: "it was not far from the city nor in a solitary part where you would still hear the sound of children playing; in this place some children got together, and *gozques* joined them; together they appeared like swarms of bees" ("no era lejos de la ciudad ni en parte tan sola que dejasen de oírlo muchachos: juntaronse tantos y con ellos tantos gozques, que parecian enjambres" [my trans.; Corominas 301]).

8 Oviedo writes that the Caribbean Indian arrow men had little dogs or *gozques* in their homes just like the disparaged *gozques* found in Spanish homes. The American *gozques* were all kinds of colors and some were shaggy and others short-haired with a silk or velvety coat. Oviedo writes that they are bit shier than *gozques* in Spain, but when they are at home they are more gregarious (3: 342).

dogs except *gozques* in the Americas (356).[9] José de Acosta writes in the *Natural and Moral History of the Indies* (*Historia natural y moral de la Indias*) that there are no "real dogs" in the Indies, just small ones: "There were no true dogs in the Indies, only an animal similar to a little dog ...." ("Verdaderos perros no los había en Indias, sino unos semejantes a perrillos ..." [231; 278]). The conquistador of Perú in the early sixteenth century, Pedro Cieza de León wrote that one sees many dogs in the Indies and he states they are similar to *gozques*, but that none of them are European breeds (Alvárez López 415).[10]

Chroniclers generally described small dogs in a parallel way to the way in which they discursively constructed Native American people.[11] Paralleling descriptions of natives, early historians highlight the indigenous American dog as feminine, a domestic pet that has no practical use value, unable to speak, and even naked. In one episode from *History of Tlaxcala* (*Historia de Tlaxcala*), Diego Muñoz Camargo's describes the Aztec sacrifice of Xolos and he fixates

[9] Garcilaso writes that in America they do not even have "pedigree dogs for hunting, such as greyhounds, beagles, retrievers, setters, pointers, spaniels, whippets, or mastiffs to guard their flocks, or even the pretty little creatures called lap dogs. There were a great many of what in Spain are called curs (*gozquillos*), large and small" ("ni aun perros de los castizos para sus cacerías, como galgos, podencos, perdigueros, perros de agua ni de muestra, ni sabuesos de traílla o monteros, ni lebreles ni aun mastines para guardar sus ganados, ni gozquillos de los muy bonicos que llaman perrillos de falda; y de los perros que llaman en España llaman gozques había muchos, grandes y chicos" [579, Riva edition 404]).

[10] The Spanish perpetuated the notion that "all American dogs were small and not pedigree." It would, however, be an over-generalization to say that dogs in American chronicles are only described as small. Sometimes, cases exist in which chroniclers describe large dogs. Oviedo tells the story of a man who had a dog who protected his peccary and would single-handedly take on and defeat jaguars (4: 388). Columbus mentions when he arrives on Fernandina Island that there were "perros mastines" (Spanish mastiff dogs). His editor, Las Casas, however, feels obliged to correct this fact and writes that if one were to really see these dogs they would not call them *mastines,* but possibly the medium-sized *podenco* (Alvárez López 415). Bernardo de Sahugún described the dogs from Mexico as a modern day anthropologist might describe them. He gathered information from native informants and combined that information with his own observations. Recent anthropological findings confirm Sahugún's descriptions. Excavations at a burial site from 650–750 Common Era in the ancient city of Tula located in Hidalgo Mexico reveal the same three types that fit Bernardo de Sahugún's sixteenth-century description of the pre-Colombian dog. The three kinds of canines include 1) a medium-sized dog with hair 2) another with similar dimensions but hairless and 3) a smaller breed with hair (about 30 per cent smaller than the medium-sized examples) (Valadez Azúa, Paredes Gudiño, and Rodríguez Galicia).

[11] The study of the Spanish description of the dog of its colonized people would merit from comparative studies with other imperial encounters with dogs. The English, upon arriving to New Zealand, also ate dogs and represented the native dogs in negative terms. One source states: "The dogs of the South Sea ... are exceedingly stupid ... and are lazy beyond measure" (qtd. in Wedde 278).

on their nakedness, calling attention three times in one sentence to their lack of hair (Beusterien, "Goya's" 167–8).[12] In contrast to the Spanish portrayal of their dogs as having the ability to communicate, Columbus began the longstanding legend that America was inhabited by silent dogs that never barked, a legend also perpetuated by Pedro Mártir de Anglería, Isabel's appointee to the Council of the Indies, and the influential chronicler Oviedo (Alvárez López).

Aside from the references to their nakedness and inability to "speak," small dogs were considered to have no purpose, a characteristic that directly tied to their association with the domestic and female space. Acosta writes that in native American houses, "if they are ill, the dog must stay by them, although they use them for nothing, only good friendship and company" ("si están malos, el perrito ha de estar allí con ellos, sin servirse de ellos para cosa, sino solo para buena amistad y compañía.") [232; 278]). Chroniclers often disparaged the dog in the American context because it was nothing more than a domestic pet. Lope de Ateizna, a Spanish cleric who observed the Audiencia of Quito in the late sixteenth century writes in his *A Historical Compendium of Peru* (*Compendio historial de los indios del Perú*):

> The ones who serve them as companions at these times are their household dogs and even the dogs of the neighborhood. They allow these to poke their snouts in the dishes, and they take their conversation and entertainment from the show that the dogs themselves provide, farting and otherwise acting like brute animals, without the diners' feeling offended or sick to their stomachs from any of this. At most they scold them and throw them out, and then at once go back to pampering them and treating them lovingly. (Salomon 120–21)

> Los que delante de sí tienen en estos tiempos por compañeros son los perros de sus casas y aun de su vecindad; a estos consienten meter el hocico en el plato, con estos tienen conversación y entretenimiento por las ocasiones que los mismos perros les dan, ventoscando y haciendo otras cosas como brutos animales, sin de cosa alguna recibir alternación ni pesadumbre en el estómago; cuando mucho les riñen y echan de allí, y sin dilación alguna, los vuelven a regalar y traer a mucho amor. (Salomon 120–21)

Attached to its small size and association with the domestic space, the *gozque* was coded as a woman's dog. The Nebrija lexicon would call the *gozque* a "catullus" (small dog) and lists a variation of the *catullus* as a "perrito de falda," a word that has semantic similarities to the English "lap dog" and essentializes the dog as belonging to a woman and her quarters. Two early seventeenth-century Spanish-French dictionaries define *gozque* as "a lady's small dog" (*petit chien de damioselle*). Girolamo Vittori defines *gozque* in Italian as a "gentlewoman's little dog" (*un piccolo cagnetto da gentil donna*) and Nicolas Mez defines it in German as "a little dog from a woman's quarters" (*Frawzimmerhundlein*) (Nieto Jiménez and Alvar Ezquerra 6: 5167).

---

[12] For more colonial sources on hairless dogs in the Americas, see Alves 109n.119.

The material and symbolic connection between the small dog, women, and Native Americans contrasted with descriptions of the Spanish dog as a powerful noble soldier. One Titian painting figuratively evokes the role of the small and large dog in the Spanish mindset of the colonial experience. Titian painted Acteon and Diana based on Ovid's *Metamorphosis* for Spain's Philip II (finished 1556–59) (Figure 3.1). Reflecting the theme of Acteon's libido found in other sixteenth-century texts, Titian depicts Acteon accompanied by a large collared hunting hound, whose angled head suggests Acteon's erect member. Diana, in turn, is accompanied by a fluffy-eared spaniel that places itself in an on guard position, barking furiously at Acteon and his dog. Beyond the sexual and gendered reading of the painting (small dog with naked female / large dog with armed male), one may also interpret it through a colonial lens. Acteon and his large dog "discover" Diana, accompanied by an African servant and a small dog.

The small dog, in contrast with the ennobled large one, had the distinction of, as the sixteenth-century progressed, becoming the first Spanish mutt. Its name suggested the growing sixteenth century social paranoia that many Spaniards were mongrels. As the modern notion of race surfaced, Gothic ancestry was one of the premier marks of purity and dictionary writers made sure to separate the *gozque* from a pure Gothic association. Francisco del Rosal writes that, because of the similarity of the sound "goz," some think that this type of dog was introduced in Spain by the Goths. But, he concludes, that is a completely erroneous assumption. The word, according to Rosal, comes from *gozo* or happiness because dogs can smile and laugh (*Alfabeto primero* 358). Covarrubias takes a different approach with respect to the *gozque* etymology, but also makes sure to separate the *gozque* from the Goth. He explains the etymology stating that the purebred Gothic origin of the *gozque* "was lost and bastardized to the extent that *gozques* are now just little dogs" ("se perdió y bastardeó, de manera que ya los gozques son unos perrillos" [my trans.; 652]). Covarrubias also connects the dog's bastardized origins with its commoner masters: "*gozques* are little dogs bred by poor and low people. They have short legs, long bodies and snouts. They annoy neighbors, bother nobles" ("que crían gente pobre y baja; son cortos de piernas, largos de cuerpo y de hocico, importunos a los vecinos, molestos a los galanes" (my trans.; 652]).[13]

In contrast to the small dog or *gozque*, the *alano* was the quintessentially big Spanish purebred. While the *gozque* and its small breed permutations became the enemy of things truly Spanish, the *alano* emerged in the seventeenth-century as the Spaniard's best canine friend. Early etymologies of *alano* contributed to making

[13] The etymology used by these dictionary writers is unsound. *Gozque* is a lexical variation on an onomatopoeia since *cusc*, *cusc*, and, especially *cucho* (this word and its variation, *chucho*, is still used in Spanish to sometimes mean "dog" or "mutt") were onomatopoeias, a common sound that people made when talking to their dog (compare with the English "shoo") (Corominas 301). *Coz* typically means a horse's kick and the word comes from the Latin *calce* (foot, heel). *Gozque* seems to incorporate this Latin root, making it "the little dog that you kick or shoo away."

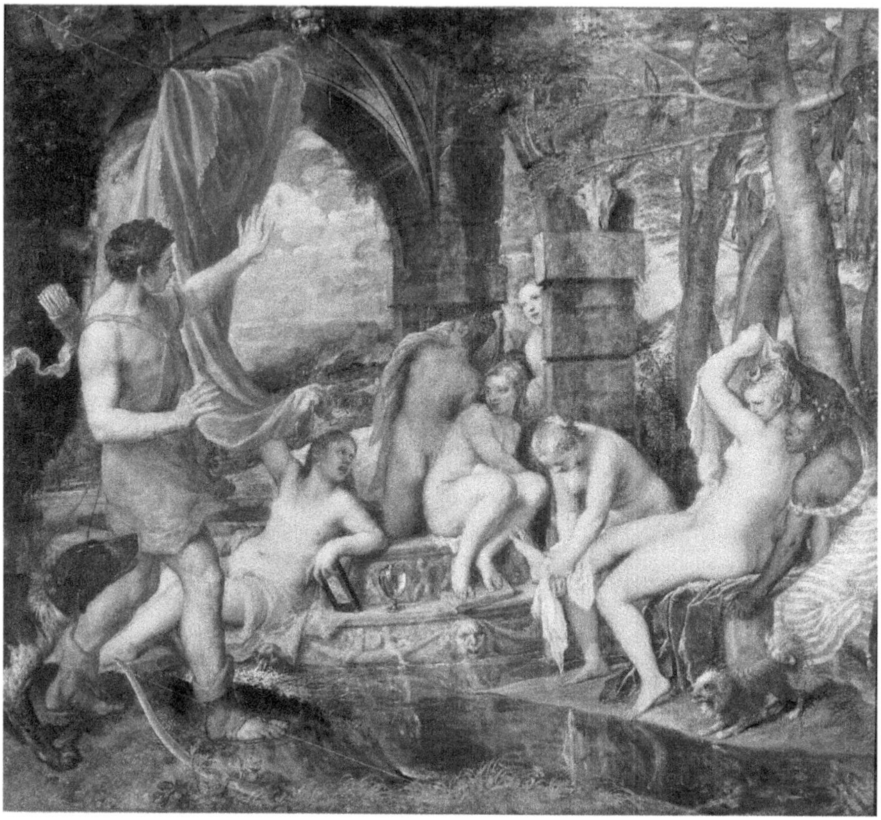

Fig. 3.1 Titian, *Diana and Acteon*, 1556–59, National Gallery of Scotland, Edinburgh, Great Britain. Photo Credit: Scala / Art Resource, New York.

it the iconic dog of the masculine Spanish self. Spanish society was pervaded by a *castizo* Old Christian Spanish craze as connected with the Gothic and the word *alano* quite literally had Gothic connotations. It meant the Alans, an Arian race from the north of Caucasus that was reputed to be fair-haired with frightening, fierce eyes. The Alans were associated with the Goths, the northern Christian tribes that had introduced Christianity to the peninsula before the arrival of the Moors. Stevens neatly gives the two definitions of *alano* in his 1706 English-Spanish dictionary: a "bulldog and also an *alan*, one of that nation" (Nieto Jiménez and Alvar Ezquerra 1: 428). The fervent Spanish desire for a Gothic identity, the primary genealogical preoccupation of the sixteenth and seventeenth century, is found in the writings of the etymology of the *alano* in which Spanish thinkers make the connection between the Gothic and the dog. Rosal for instance comments: "*alano*, a dog belonging to the Alans, the nation that introduced it to Spain" ("*alano*, perro de los *alanos*, nación que los introdujo en España" [my trans.; Nieto Jiménez and Alvar Ezquerra

1: 428]). The purity of this animal would form part of a popular proto nationalist celebration of the Spanish body in which the Gothic animal became a watchdog against the heathen.[14] A popular poem from the period that describes a dog from Alba, renowned for its ability to smell out the Jew, labels the dog as an *alano* (l. 60, Gillet 422; also see Beusterien, *An Eye* 119–120).

Testifying to its significance in the hierarchy of early modern breeds, the oldest poem in Spanish dedicated to an animal breed or *casta* is the seventeenth-century anonymous poem "El Alano" (1604).[15] "El Alano" is not just the first poem written in Spain dedicated to a dog breed, but it is the first poem in European history dedicated to an animal breed. The *alano* was considered the largest of Spanish dog breeds in the early modern period. In the poem, as "lion" of Spanish dogs, the dog's name is León. In "El Alano," León contrasts with the typical commoner's dog, the barking *gozque.* The small size of the *gozque* in the poem, in contrast to the large *alano,* is highlighted further by the Spanish suffix *zuelo* that serves to make the noun diminutive.

> The thousands of vile mutts
> that run through the streets
> envy his strength and elegance.
> They follow him with their tails
> between their legs.
> They approach him barking,
> but he does not even acknowledge their presence. (my trans.)

[14] Some canine scholars today sometimes still follow these early modern examples and assume that the *alano* was brought to Spain by the Alans. Although the word *alano* is found to describe dogs in medieval Europe, the shape of the large hunting dog in Spain most directly comes from an Arabic breeding tradition. One medieval Arabic-Spanish hunting manual describes a corpulent dog with a strong, large head, pug-nosed with a short, smooth coat of hair. *The Book of Hunting Animals* (*Libro de los animales que cazan*) ('Abd Allah), a thirteenth-century Spanish translation of a ninth-century Arabic hunting treatise provides the first and earliest detailed description of the large hunting dog's face and head. The first three *tratados* of the Arabic text, often called *Moamín's Book* (*El libro de Moamín*), discussed falconry and the last two discussed hunting dogs. The Arabic text would be a direct source for medieval hunting texts, most especially, Alfonso XI's fourteenth-century *Hunting Book* (*Libro de la montería*) (see Alfonso XI). According to the medieval Spanish translation from the Arabic, ideal dogs (it does not mention the word *alano*, but simply "dog" or *can*) should have "light heads and wide and tall foreheads, and floppy, skinny, and long ears with a big space in between each ... and there should be a lot of veins in their forehead" ("las cabeças ligeras, e las fruentes anchas e altas, e las orejas floxas e delgadas e luengas e que ayan grand espacio entrel una e la otra ... e que ayan muchas venas en las fruente" [my trans.; 'Abd Allah 220-21]).

[15] All quotations are taken from an early twentieth century edition ("*El Alano*"). The poem was originally published in a book that discussed a variety of different household bird varieties. It can be found in Juan Bautista Jamarra's "A Guide to Ten Small Cage Birds" ("Conocimiento de las diez aves menores de jaula") published in Madrid by the Royal Press in 1604.

> Que era mirarle calles discurriendo,
> mil gozquezuelos viles, envidiando
> tanta robustidad, y gala tanta,
> siguen medrosos, cercanle ladrando,
> mas ni hace caso dellos… (14)

The poet praises León for his size, the vice like grip of his jaws, his success in the hunt, as war soldier, and in bringing down the bull. As a memorial and epitaph to his beloved large dog, the poem contrasts the later Spanish tradition of mock epitaphs that were written for toy dogs belonging to women.[16] Poems such as "El Alano" are the culmination of the deployment of the *alano* for expressing a specific character, specifically as a uniquely national breed connected to masculine Spanish strength. These qualities infused early notions of the human as the poet describes the *alano* as "terrible and fierce, but playful and human" ("terrible y fiero, retozón y humano" [my trans.; 14]). In an anthropocentric logic, the small dog, like its master, became a lesser human, while the large dog became more human.

In the prologue to Part II of *Don Quijote,* Cervantes discards the large-small dog anthropocentric tradition. Cervantes makes reference to both the large and small breed when he describes a madman crashing stones upon the heads of dogs. After the madman is punished by a dog owner for having hit a *podenco*, the madman does not hit any more dogs, including *alanos* and *gozques*: "In fact, all the dogs he encountered, even if they were *alanos* or *gozques*, he called *podencos*, and so he never dropped a stone" ("En efeto; todos cuantos perros topaba, aunque fuesen alanos, o gozques, decía que eran podencos; y así, no soltó más el canto" [457; 620]). Referring to dogs that are neither big like the *alano* nor small like the *gozque*, Cervantes evokes and steps outside of the big–small dog discursive tradition when he creates the medium-sized dog protagonist in the prologue.

A symbolic anthropocentric tradition invested the large and small dog with aspects of human character since ancient times. The tradition was articulated in a specific way in the Spanish context, especially after the conquest of America. In turn, no consistent discursive paradigm or significant literary attention was attached to the medium-sized dog. In constrast to the firmly established *alano* and *gozque* literary and cultural tradition, the anthropocentrizing valence given to the *podenco* was ambiguous and Cervantes thereby created a new literary purpose for the dog when he created the *podenco* in the prologue.

Popular sayings express admiration for the *podenco*, in contrast to the useless small dog: "No, it's a *podenco* so you should not kill it or harm it. It is a useful dog" ("No, que es podenco. Que no se mate ni haga mal, porque es perro de provecho" [my trans.; Molho]). Cervantes uses a version of this expression, as Maurice Molho writes, when the haberdasher defends his *podenco.* The *podenco* was a hunting dog and some sources ennobled it like the *galgo* (*Diccionario* 136).

[16] In Spain, one poet writes a mock poetic epitaph for Jasmine (Beusterien, "Goya's" 162). In New Spain, another poet writes mock funerary honors to Pamela (Tortorici).

Francisco del Rosal writes that the *podenco* is the same as a *galgo*. One popular ballad unites both dogs together. In the "The Oath of Saint Gadea" ("Jura de Santa Gadea") the Cid goes into exile and takes his falcons with him but leaves behind his chained *podencos* and greyhounds (Pidal).

As opposed to the *alanos* in "El coloquio de los perros," Cervantes chooses a *podenco.* Even though the *podenco* had some large dog characteristics (use value as a hunter), sixteenth-century hunting manuals clearly distinguished each breed and, indeed, did not treat the *podenco* with the same importance given to other large hunting dogs like the *galgo* and the *alano.* Although some sources described the *podenco* as a "big" dog and no different from the *galgo,* other sources made a clear distinction. Luis Barahona de Soto's *Hunting Dialogues* (*Diálogos de la Montería*) gives the most complete sixteenth-century description of the *podenco,* clearly distinguishing it from the *galgo* both in its physical description and purpose in the hunt.[17] One hunting manual does not even describe the *podenco* because it is not a noble breed. In the first published hunting manual, *Discourses on Hunting* (*Discursos de montería*), Argote de Molina writes that he only wishes to describe big game hunting done by noblemen—*caballeros* (gentlemen) with

[17] Contemporary Spanish dog manuals generally distinguish between three *podenco* types, the Andalusian, Canary and Ibicenco and some also identify a Portuguese and Galician *podenco* variety. They describe the *podenco* as "part *galgo*" since it hunts in the same way as the greyhound, using its sight but also smell and hearing (Benito Ruiz and Malo Acrudo). Recorded descriptions of the *podenco,* however, already can be found in the sixteenth century. In his hunting dialogue, Soto describes its color as reddish and gives a description quite similar to what would be known today as a large-type Italian greyhound. *Podencos* chase the rabbit "better than the wind" and their diligence in tracking the rabbit is exceptional (*Diálogos*; Barahonda). Here are the Soto's conclusions on the *podenco*: "A good *podenco* should have five defining characteristics: it can sniff out the prey, return to sniff it out, take it away, retrieve it, and find its lair. 'Sniffing out' means they uncover all sorts of prey because of their great stamina, scent and spirit. 'Returning to sniff out' means that if they lose their prey in a high, thickly-forested mountain, they will return again and again until they have cornered it. 'Taking it away' means that they are able to get the prey before any other dog in whose company they may be. They 'retrieve' it, bringing it to their master night and day. They even bring it home on their own since many times the *podenco* is sent off to hunt on its own. In the summertime rabbits stay in their cool warrens all day because of the heat and other rabbits take refuge hidden in the mountain. 'Find its lair' means that the *podenco* knows how to find the rabbit's hiding place without getting exhausted and without fruitlessly searching all day" ("En el buen podenco deben concurrir cinco partes principales: que halle y rehalle y quite y traiga y llame de morada; que halle mucha caza por su gran diligencia de pies y viento y codicia; rehalle, porque si la perdiere por ser el monte alto y espeso, la torne a hallar muchas veces hasta encerrarla; que quite a otros cualesquier perros en cuya compañía se hallare y la traiga al cazador de noche y de día, y aun a casa si el se saliere solo a caza como suele suceder muchas veces; y llame de morada, porque en verano suelen muchas veces estar los conejos cerrados todo el dia en los vivares frescos por el calor del sol y otras por falta de monte, y el perro que no lo hace cansase de andar buscando todo el dia y no puede topar ninguno" [my trans.; 476–7]).

caballos (horses)—and he is not concerned with the hunting by commoners that includes rabbit hunting with *podencos* (91).[18]

To further underline the ambiguous valence of the *podenco,* Cervantes pairs the dog up with a haberdasher. Just as the *podenco* was often looked down upon as a commoner's breed, but it was still a special dog, so the hat profession was at once considered that of a commoner but also as having edifying, ennobling connotations. Before their expulsion, elite *morisco* families (such as the Lakhousa family, descendants of the Abencerrajes), specialized in hat-making, especially in Toledo (Altosano). The *bonete* (hat) was a sign of noble status and the haberdashers that made *bonetes* also had elite status. Nonetheless, writers criticized the profession because it was a common job and not the occupation of noblemen. Sometimes the jibe against the haberdasher was not just that it was practiced by commoners, but those of non-Christians, such as Moors and Jews. One popular ballad described it is a respectable profession, but undercuts it by calling it a "Jewish job" (Cid 425).

Aside from the big–small dog dichotomy, the humorous animal spectacle was another example of anthropocentrism that existed when Cervantes set out to write two dog anecdotes in his prologue. Medieval, Old Regime attitudes placed the animal at the center of popular ceremonies. In observing animal abuse in public spectacle, the privileged and underprivileged projected themselves in the position of animal *and* human. The momentary feeling like the abused animal *and* also feeling like the one who inflicts pain causes laughter, loud Rabelaisian laughter, as the spectator associates and disassociates himself from the animal.

The animal spectacle did not disappear after the medieval period. In *The Great Cat Massacre*, Robert Darnton examines a moment of cat abuse that he sees as crucial to his translation work as historian. Two apprentices at a printing press in the late 1730s in France smashed all the cats' spines that they could find and then staged a mock trial pronouncing the animals guilty and stringing them up on an improvised gallows. The event is described as an absolutely hilarious event for the apprentices and the onlookers of the event are reported to have broken out in uproarious laughter. By understanding the humor of the cat massacre, Darnton argues, it is possible to better understand how artisanal culture functioned in the Old Regime. In this case the humor, the spectator's association and disassociation with the animal and master, connects to the animal conditions in which the apprentices live (in contrast to their masters who treat their domestic cats much

---

[18] Argote de Molina published the first hunting manual in Spanish in Seville in 1582. The social significance of the publication of his hunting manual is enormous since it marks the first time a book was published on hunting for an audience that was not exclusively the elite, whether a noble family or a king. Gonzalo Argote de Molina edited and amplified medieval hunting treatises, but his was revolutionary in the sense that it was not only a manuscript written for the aristocrat, but a published book for the public at large. The availability of details of breeding hunting dogs in printed form as opposed to the earlier manuscript tradition signals an emerging modern class interested in reading about the activity of hunting and animal breeding, formerly a field of knowledge under the exclusive purview of royals and nobles.

better) as well as the sexual charge of the cat scene (the cats represent the sexuality of the master's wife).

The animal spectacle in Cervantes' time was widespread and functioned as the center of merriment for the carnival or town feast days. In the cat scene from part II of *Don Quijote*, the duchess entertains herself and her court by having a bag of cats with bells tied to their feet thrown in Quijote's quarters.[19] In an act of playful cruelty, some boys lift up the tails of Rocinante and Sancho's donkey and place prickly furze in each's rear.[20] Dog performances also seem to have been quite popular.[21] In early modern Europe and New Spain, one finds carnival-type celebrations of dog weddings and baptisms (Tortorici). In Part I of *Don Quijote* Cervantes compares the way Sancho is made fun of when he is blanketed with the same action done to a dog: "And there, with Sancho in the middle of the blanket, they began to toss him and make merry with him as if he were a dog at Carnaval" ("Y allí, puesto Sancho en mitad de la manta, comenzaron a levantarle en alto y a holgarse con él, como con perro por carnestolendas" [122; 170]).[22] Although contemporary readers may not laugh at Sancho being tossed at a blanket, Cervantes describes the event as one that enormously embarrassed Sancho, causing great merriment and laughter for its spectators. In Spain, all sorts of animal spectacles continued well beyond the early modern period that once had a similar comic purpose. Manganeses, a small town in Spain, only officially eliminated the throwing a goat from a tower during the town feast day at the beginning of the twentieth-first century.

Cervantes was well aware of the humorous tradition of animal abuse when he set out to write his prologue. The supposedly comical story of a crazy man slamming a dog with a rock appears in Correas' compilation of proverbial expressions, showing the popularity of a rather gruesome dog story in the early

---

19 Spanish Golden Age cats need further study. One may argue that just as Velázquez increases focus on the dog in *Las meninas*, so he also does so for the cat in the foreground of *The Fable of Arachne* (*Las hilanderas*) and thereby pushes his viewer to consider the animal and its unique otherness. In literature, one finds many examples of cat descriptions, including this scene from Cervantes, María de Zayas' novella "The Miser's Reward" ("El castigo de la miseria") (for comments on the tortured cat in this piece, see Greer, "María") and, most famously, Lope de Vega's *Catfight* (*Gatomaquia*).

20 Alves points out that Cervantes condemns this act of animal abuse by pointing out that the narrator calls the boys "worse than the devil himself" ("mas malos que el malo" [qtd. in 59]).

21 Covarrubias mentions dog shows and Berganza also describes an episode when he had been used for a show. In one source text for "The Dialogue of the Dogs," Alcázar's *pícaro* dog Zarpilla recounts a dog stunt episode in which he was the protagonist: "I jumped up in the air and / I hopped about to entertain him" ("daba saltos en el aire, / triscaba por complacelle" [my trans.; 428]). One can find medieval images in Spain and across Europe that describe dog shows, some of which are identical to Berganza's trick in which the person holds up a ring and the dog jumps through (Martin 1568).

22 Compare with the same usage of the expression in the picaresque novel *Guzmán de Alfarache* (359).

modern period. Another version of the story is repeated in Melchor de Santa Cruz's *Spanish Anthology of Apothegms or Wise Sayings, Humorously Recorded from Some Spaniards* (*Floresta española de apotegmas, o sentencias sabias y graciosamente dichas de algunos españoles*). Melchor de Santa Cruz describes a man who was bitten by a dog and then who took a stone in hand and bashed it on the sleeping dog's head warning the dog to not sleep while his enemies lived (363). Melchor de Santa Cruz includes the story of the dog and the madman in a list of jokes about madmen with popular stories as the source for his jokes, as the title of his book makes clear. Indeed, Melchor de Santa Cruz's book, which first appeared in Toledo in 1574, turned out to be quite popular and three new editions of the book would appear two years later in Zaragoza, Salamanca and Alcalá.

The Melchor de Santa Cruz story draws on a tradition that finds humor in the physical abuse of the animal, a deep-rooted anthropocentric tradition that existed in the animal spectacle in Europe from the ancient to the medieval period to modern times. In his prologue, Cervantes rejects this tradition and its humor when he asks his reader to be sympathetic to the dog rather than laughing at its plight. Despite the continued existence of the humorous animal spectacle, the early modern period was increasingly more sensitive and humane toward the animal. One of the first active proponents of animal rights, William Hogarth, criticized carnival-type animal abuses that served entertainment purposes such as in the print entitled "First Stage of Cruelty" in which Hogarth lambasts all sorts of such abuses, including a group of boys poking an arrow up the anus of a dog (see Shesgreen, Plate 77 for this illustration).

In Spain, one early example of an expression of a humane attitude toward the animal can be found in another version of the mad-man story who hit dogs on the head with a stone. Mateo Alemán had recorded his version of the story in 1604 in the second part of the picaresque novel *Guzmán de Alfarache*. Alemán writes that the crazy man took revenge on a dog for having bit him. Alemán, however, does not make the primary purpose of the story either humorous nor carnivalesque. Alemán is a precursor to the humane as it means to act humanely toward animals when he writes that it serves as an exemplary tale for not being vengeful toward one's neighbor. Alemán remembers the gospel story of loving one's neighbor and equates the body of the dog with the body of one's human neighbor. The story of the dog in Alemán's novel is an explicit moral lesson. By making the "other" a dog, Alemán urges people to exercise compassion for one another, both humans and animals.[23]

Although he urges the reader to be sympathetic to the *podenco*, Cervantes does not follow Mateo Alemán who directly condemns the madman's act of animal abuse with a didactic, humane lesson. In other words, Cervantes does not create an animal that should be read through the logic of the human face. Rather than making it human-like, Cervantes gives the dog the shape of a book, more specifically *his* book, *Don Quijote.* Cervantes uses the "Prologue to the Reader" for Part II, to

---

[23] See Thomas 173 for parallel denunciations of animal cruelty in early modern England.

justify the writing of his book. The two dog anecdotes in the prologue directly link the character of a madman with Alonso Fernández de Avellaneda, the author who wrote a 1614 unauthorized and spurious continuation of *Don Quijote*. The two dog stories in the Prologue of Part II respond to Avellaneda since Cervantes tells his reader how the apocryphal author Avellaneda has abused the original.

The first story in the Prologue concerns a crazy man who sticks a tube made out of reed into a dog's anus and blows up the dog.

> In Seville there was a madman who had the strangest most comical notion that any madman ever had. What he did was to make a tube out of a reed that he sharpened at one end, and then he would catch a dog on the street, or somewhere else, hold down one of its hind legs with his foot, lift the other with his hand, fit the tube into the right place, and blow until he had made the animal as round as a ball. (456)

> Había en Sevilla un loco que dio en el más gracioso disparate y tema que dio loco en el mundo. Y fue que hizo un cañuto de caña puntiagudo en el fin, y en cogiendo algún perro en la calle, o en cualquiera otra parte, con el un pie cogía el suyo, y el otro le alzaba con la mano, y como mejor podía le acomodaba el cañuto en la parte que, soplándole, le ponía redondo como una pelota. (619)

Cervantes tells this story because he wants his reader to know that Avellaneda's sequel has taken his "dog," the first part of *Don Quijote*, and blown it up out of proportion, exaggerating the character of Don Quijote, for example, in unsightly ways.

This first dog anecdote concludes with two rhetorical questions. The madman blows up the dog and turns to his onlookers, stating: "Now do your graces think it's an easy job to blow up a dog?" ("¿Pensarán vuestras mercedes ahora que es poco trabajo hinchar un perro?" [457; 619]). In the sentence following the madman's rhetorical question, Cervantes breaks out of story mode and the narrator addresses the madman directly as if he were Avellaneda: "Now does your grace think it's an easy job to write a book?" ("¿Pensará vuestra merced ahora que es poco trabajo hacer un libro?" (457; 619]). Although he does not specifically state the name Avellaneda, Cervantes addresses Avellaneda because he begins the anecdote telling his reader (the prologue is addressed to a reader whom Cervantes addresses in the *tú* form) that he wants his reader to tell Avellaneda the anecdote about the dog. The answer to the first rhetorical question posed by the madman is obviously "no." It does not take a lot of work to blow up a dog. Only an insensitive foolish person is needed. Cervantes implies that Avellaneda did not work hard to blow up his version of the Quijote character. The answer to the second question posed by the authorial voice and directed to Avellaneda is obviously "yes." It was not "*poco trabajo*," but a considerable effort: Cervantes worked hard on the first (and second) volumes of his book. With the second rhetorical question, Cervantes makes Avellaneda a madman who foolishly thinks blowing up a dog is a lot of work.

The dog as figure for Cervantes' original *Don Quijote* acquires currency after the dog's body is read literally. Many readers do not know what to make out of the story of blowing up the dog: they write that the madman that blows up the dog is an allegory of something else. Indeed, the word "dog" in the contemporary expression *soplar un perro* (blowing up a dog) in Spanish does not refer to a dog. In modern Spanish parlance, "dog" is only a figure (and not a dog) because "blowing up a dog" is a clichéd expression in which "dog" is proverbial (one may compare the meaning of dog in the expression with the meaning of "dog" in the English expression "to beat a dead dog"). But in Cervantes' Spanish, the dog *was* a dog. It is important to note that the anecdote in the prologue to Don Quijote is the origin of the contemporary expression. In other words, the story is new with Cervantes and, for the first readers of the text, the body of the dog in this anecdote is a real body of a dog. As it was in Hogarth's England, the incident of dog blowing was probably known and experienced by many of Cervantes's readers. Instead of employing carnivalesque laughter *at* the dog, Cervantes, like Hogarth, urges a sympathetic reading toward the animal that has been literally distorted out of shape and blown out of proportions. But, after the literal reading that sympathizes with the animal, the linking of the dog's body with a book functions as a more effective, persuasive literary trope since the reader has felt for the animal.

The dog suffers and is a book. In the prologue, Cervantes did *not* connect the dog with himself as author even though there is evidence in his writing, both textual and biographical, that suggests at times he associated himself as author with a dog. Rather than a dog–author connection, the prologue connects the dog and the physical object of the book. In this way, it evokes the book burning scene from Part I. Like the living dog that is also a book, the anecdote pushes the reader to imagine the books in Don Quijote's library as living bodies that are not only human, but also animal.

In contrast to the first anecdote, the second dog anecdote draws from literary and popular sources, specifically those that tell of a madman who hits dogs with a rock. Cervantes continues the prologue stating that if the first dog anecdote is not clear, then he has one more anecdote that he wants his reader to tell Avellaneda. In this anecdote, Cervantes introduces the *podenco*. Like the first anecdote, the second story in the prologue also associates Cervantes' book with the body of a dog. The one who bothers the dog is also a madman. In the second anecdote the madman is from Córdoba, not Seville.

> In Córdoba there was a madman who was in the habit of carrying on his head a slab of marble, or a stone of no small weight, and when he came across an unwary dog, he would go up to the animal and drop the weight straight down on it. The dog would go into a panic and, barking and howling run up three streets without stopping. (456–7)

> Había en Córdoba otro loco, que tenía por costumbre de traer encima de la cabeza un pedazo de losa de mármol, o un canto no muy liviano, y en topando algún perro descuidado, se le ponía junto, y a plomo dejaba caer sobre él el peso. Amohinábase el perro, y, dando ladridos y aullidos, no paraba en tres calles. (619)

The second madman takes a stone and smashes it on sleeping dogs. They leap to their feet and run off spooked and howling.

Cervantes directly associates the violent action of the madman with what he perceives as the violent action that Avellaneda has taken against his *Don Quijote*. Cervantes's dog, his book, is bashed with a stone. The stone, in turn, that strikes the dog is Avellaneda's book. Cervantes makes explicit the connection between stones and Avellaneda's book when he writes that letting loose the rock on the dog was the same as that author that let loose his wit "in books, which, when they are bad, are harder than boulders" ("en libros que, en siendo malos, son más duros que las peñas" [457; 619]). The heavy tomes of Avellaneda lack wit, grace, imagination and are harder than rocks. Avellaneda's *petrolos*, his heavy rocks, his insensitive tomes, ruthlessly smash on innocent resting dogs, original published versions of Cervantes's *Don Quijote*. Cervantes created a living dog. The dog like his book slumbers until a person comes along and, like the reader of a book, decides to use it for a specific purpose or just enjoy its company. The madman author will not let the slumbering dog lie.

The solution to the problem of the perverse madman who enjoys seeing dogs suffer is resolved by a dog lover who disciplines the madman with a canine vocabulary lesson. That dog lover beats the madman and with each strike repeats the name of a special kind of dog: "el nombre de *podenco*." Cervantes concludes the second anecdote as follows:

> Now, one of the dogs he dropped the weight on happened to belong to a haberdasher and was much loved by its owner. The stone came down, hit the dog on the head, and the battered animal began to yelp and howl; its master saw and heard this, and he seized a measuring stick, came after the madman, and beat him to within an inch of his life, and with each blow he said:
>
> "You miserable thief, you dog, why did you hurt my *podenco*? Didn't you see, cruel man, that my dog was a *podenco*?"
>
> And repeating the word *podenco* over and over again, he beat and pummeled the madman. Chastised, the madman withdrew and was not seen on the street for more than a month.[24] (457)

[24] I have modified the translation, changing Grossman's "hound" to *podenco.* In this book I generally avoid translating Spanish dog breeds like *alano* to English because of the specific national inflection the name and because it makes no sense to impose an English post-nineteenth-century classification on to the seventeenth century. The history of the translation of the word *podenco* to English provides a case study of the difficulty of translation. *Podenco* has had many variant translations into English since the late sixteenth century. English hunting manuals contemporary to the *Don Quijote* generally translated *podenco* as a "bloodhound." Three Spanish-English lexicons define *podenco* as follows: "a dog called a blood-hounde, sagax, indagarius" (Richard Percivale, 1591); "a blood hound to smel out" (Richard Stepney, 1591); and "a blood-hound, a tumbler" (Minsheu, 1599) (Nieto Jiménez and Alvar Ezquerra 8:7781). Translations of *Don Quijote* never used "bloodhound" for *podenco* even though it was the favored translation in early Spanish-English dictionaries. Instead, "spaniel" would be the only translation of *podenco*

> Sucedió, pues, que entre los perros que descargó la carga fue un perro de un bonetero, a quien quería mucho su dueño. Bajó el canto, diole en la cabeza, alzó el grito el molido perro, violo y sintiólo su amo, asió de una vara de medir, y salió al loco, y no le dejó hueso sano; y cada palo que le daba decía:
>
> —Perro ladrón, ¿a mi podenco? ¿No viste, cruel, que era podenco mi perro? Y repitiéndole el nombre de *podenco* muchas veces, envió al loco hecho una alheña. Escarmentó el loco y retiróse, y en más de un mes no salió a la plaza. (619–20)

The haberdasher punishes the madman with his measuring stick and repeats the word *podenco*. The purpose of the yardstick, or the *vara de medir*, is measuring and it evokes the out-of-bounds act of the first madman who had blown something beyond its proportions.

The *podenco*, like Cervantes's book, should not be damaged. Moreover, all people, even commoners, should have access to the dog-book. The *podenco* was nearly a universally known dog. Soto writes in his hunting manual that a description of the *podenco* is unnecessary because it is so well known (although he still provides one) (476). Cervantes is the first author to take the popular story of the dog that was smashed by a madman's rock and to then insert the figure of the *podenco*, making the dog a substitute for his book. Cervantes, with the *podenco*, offers an authorial gift to his reader: the *podenco* is a book-dog and Cervantes understands his book as dog in the sense that both are alive and provide emotion and humor.

The second man drops heavy stones on the sleeping dogs and Cervantes intends to right the wrong with his prologue. The main way that Cervantes rights the wrong is by writing his own sequel, Part II of *Don Quijote*. But the most immediate way that he rights (and writes) the wrong is by teaching the madman the meaning of *podenco*. The vocabulary lesson worked: when the madman returned, with even more evil intention to hurt lying dogs, he did not smash his rocks on any dog for fear the dog might be a *podenco*.

---

for over two hundred years, showing that the English had already favored the "spaniel" as a special friend and an early English breed. Five years after the publication of part II of *Don Quijote* in Spain, Thomas Shelton translated *podenco* as "spaniel" for his 1620 English translation. Aside from Shelton, the other early popular translators of *Don Quijote*—Shelton, Jarvas, Motteux, and Smollet—translate *podenco* as "spaniel." With the advent of nineteenth-century breed consciousness, however, English versions of *Don Quijote*, since the nineteenth century, do not agree on the translation of *podenco*. Breeds were not standardized until 1873 so translators had attempted to capture the spirit of the *podenco* with "spaniel." But after 1873, translators do not use "spaniel" any more for *podenco* and try to portray how they believed the animal looked (like a lurcher, whippet, and greyhound) or to simply suggest that it had a hunting function (like a pointer or a hound). The following translations of *podenco* appear in English editions of *Don Quijote* after 1873: "lurcher" (Ormsby, 1885), "pointer" (Watts, 1898 and Lathrop, 2005), "greyhound" (Putnam, 1949), "whippet" (Raffel, 1999 and Rutherford, 2000), and, finally, "hound" (Grossman, 2003).

> He would go up to a dog, stare at long and hard, and not wanting or daring to drop the stone, he would say:
>
> "This is a *podenco*: watch out!"
>
> In fact, all the dogs he encountered, even if they were *alanos* or *gozques*, he called *podencos*, and so he never dropped a stone. (457)
>
> Llegábase donde estaba el perro, y mirándole muy bien de hito en hito, y sin querer ni atreverse a descargar a piedra, decía: —Este es podenco: ¡guarda!
>
> En efeto; todos cuantos perros topaba, aunque fuesen alanos, o gozques, decía que eran podencos; y así, no soltó más el canto. (620)

When he sees a dog, any dog, be it the small lap dog runt, a *gozque*, or the great powerful *alano*, the madman is haunted by the word *podenco*. Upon coming across any dog, he connects the dog literally to a new sign of *podenco* and will not separate it from the beating by the haberdasher. In the creation of this dog that is neither large nor small, the prologue, and the exercise of writing Part II of *Don Quijote*, to a large degree, Cervantes's warns Avellaneda and similar-minded followers to keep away from the dog. Cervantes uses the anecdote to say: "do not come close to my dog, my canine text, with your hot air or your stones. The *podenco* comes to life and wags its tale when one who comes along and wants to experience it."

What is, if any, the usefulness of the *podenco* or the name *podenco*, not for the haberdasher, but for the reader? The *podenco* anecdote is used to communicate the idea that Alonso Fernández de Avellaneda should not have treaded upon the original *Don Quijote* since he distorted it (into a blown up dog) and put dead weight (*petrolo*) on it. Cervantes suggests that Avellaneda did not have the spirit (breath) that came from Cervantes's lungs. No breath from Avellaneda entered the Quijote body. Avellaneda only brought words to a clichéd, lifeless state. Avellaneda's book was not light and free to roam, but a rock, heavy and dead.

With the two dog anecdotes, Cervantes makes it clear to his reader that he disapproves of Avellaneda's book. Nonetheless, beyond his disapproval, Cervantes' *podenco* anecdote also raises a larger question about the role of human sympathy as it is expressed toward other humans, animals and material objects (in this case, a book). The story of the *podenco* stimulates the following question: to what degree does a human-animal hierarchy inform and determine moral responses? After he hits his dog, the haberdasher lambasts the madman, saying "you miserable thief, you dog" (*perro ladrón*). In this instance, the haberdasher makes the madman an animal, drawing on the position that associates brutish behavior with dogs. For the haberdasher, as in the version of the story told by Alemán, the madman is wrong for not being sympathetic toward the dog. But the haberdasher's expression of "you dog" is directed against a person, forcing the reader to laugh, but also to question the human-versus-animal distinction itself. Although the haberdasher calls another man a dog, is it not obvious that the *podenco*, not the madman, is *really* the dog?

As mentioned in the previous chapter, the reader's sympathies were toward Berganza in "The Dialogue of the Dogs" until he calls a human a dog. In the prologue, as opposed to Berganza calling a person a dog, a human *person* (the haberdasher) calls another person a dog, hoping to inspire sympathy for his real dog. Sympathies were toward the *podenco*, but the nature of that sympathy is called into question, like the Berganza case, when a human is called a dog. The reader is pushed to question the moral logic behind expressions that animalize humans and, indeed, the role that the animal has in how humans define themselves as human. Who is the person to whom sympathies should lie—an animal person, a human person, or a book person?

Ultimately, the story of the *podenco* stimulates the question: what should be the shape of sympathy toward animals and objects such as the book? Mateo Alemán told the dog story that served as an allegory as a human story: it is a story about the particularity and generality that constitute human sympathy as Immanuel Levinas describes it. Levinas asserts that "the ethical extends to all living beings" because the human makes manifest the life of the other that is "more important than my life." We sympathize with the particular (the human individual) and draw from the particularity a universal ethics in which we sympathize with the general (all beings). But through the dog-book motif, Cervantes redirects the object toward whom ethical decisions are directed. In other words, Cervantes does not ascribe to the Levinas imperative toward the human since he questions what constitutes the human component behind compassion and the ethical imperative that drives it.

The haberdasher may have a *podenco* for his friend. But for the reader the *podenco* is not a friend. It is a dog-book. That dog-book does not serve, like Alemán's version of the story, to communicate a humane moral or as a figure for human friendship. The newly-crafted *podenco*—like the paradoxically large-small *alano*s in the dog dialogue and the Spanish mastiff in *Las meninas*—are not friends in the anthropocentric sense. They signal that the human does not hold a monopoly on sympathy (although its place is not forgotten). The dog-book occupies a place in which an unknowable, infinite amount of non-human persons have faces that demand responsibility.

Chapter 4
As Death Approaches: The Dog in *Las meninas*

The dog as a threshold figure separating the living and the dead has a long history. In the classical world, Hecate, the goddess of threshold spaces and death, was accompanied by a dog or dogs. Also in the classical tradition, the canine figure of Cerberus guarded the entrance between the living world and land of the dead. The tradition of the dog as a threshold image does not only exist in Greek and Roman mythology. In the Indian tradition, the lord of death, Yama, was accompanied by a pair of four-eyed watchdogs, Syama and Sabala, offspring of Lord Indra's bitch Sarama. The Aztec Xolotl, often depicted as a dog-headed man, was a soul guide, a similar figure to the Egyptian canine-like lord of death, Anubis. A guard dog also keeps watch over the entrance to the next world in myths from the Middle East, Iceland, the Finns, and America. The Taino people of Hispaniola believed that a gate-master dog sat at the edge of a lake, controlling access both in and out of the dark realm of the nonliving (Schwartz 94).

A North American Native origin story also posits the dog as a liminal figure. It begins by describing a scenario where animals had emerged coevally with humans (in contrast to Freud and Hobbes' portrayal of the animal as chronologically anterior to the human). This Ojibwa legend explains that in the beginning the earth trembled and a great rift appeared, separating the first man and woman from the rest of the animal kingdom. As the chasm grew deeper and wider, all animals, afraid for their lives, disappeared into the forest—except for the dog, who after much consideration leapt the perilous rift to stay with the humans on the other side. His love for humanity was greater than his bond with other creatures, he explained, and he willingly forfeited his place in the forest paradise (Thurston).

The dog in the Ojibwa story crosses the space that separates humans and animals and it knows what it is to have lived with both. Virginia Woolf's *Mrs. Dalloway* describes a dog that bridges the separating abyss between animals and humans in another way: the dog begins to transform into a human. Septimus Smith, the World War I veteran, suffering from shell-shock, becomes hysterical when a dog approaches him and morphs into a man:

> No crime; love; he repeated, fumbling for his card and pencil, when a Skye terrier snuffed his trousers and he started in an agony of fear. It was turning into a man! He could not watch it happen! It was horrible, terrible to see a dog become a man! At once the dog trotted away.
>
> Heaven was divinely merciful, infinitely benignant. It spared him, pardoned his weakness. (51)

The character Septimus Smith survived a war that surpassed all previous ones in the sense that soldiers saw so many die at extremely close range. When he returned from that war, he sees a dog transform into a man before his eyes. Were he to have seen the dog completely changed into a human, he would have seen all that horrible death once again. But heaven is good to Septimus because he does not have to face mortality. The dog trots away and Septimus' experience with the horrible dog-almost-human is over.

Giorgio Agamben's *The Open: Man and Animal* offers a way to interpret the abyss separating the human and animal in the Ojibwa story and the momentary man-dog hybrid in the Septimus story. Bare life for Agamben is the chiasmic interval between animal and the human that is neither animal nor human. The hybrid is a synonymous metaphor for the bare life because it functions as a conceptual threshold between the human and animal, and between life and death. To begin *The Open*, Agamben describes illustrated human forms with animal heads that depict the righteous at the apocalypse. He returns to the same metamorphic images again for the conclusion of *The Open*: "Perhaps there is still a way in which living beings can sit at the messianic banquet of the righteous without taking on a historical task and without setting the anthropological machine into action" (92). Only through bare life can theory and practice be realized at a "messianic banquet of the righteous" outside of the combined straightjacket of the historical task and the anthropocentric machine. Agamben insists on the bare life in order to arrive at a new politics: "Until a new politics—that is, a politics no longer founded on the *exceptio* of naked life—is at hand, every theory and every praxis will remain imprisoned and immobile" (*Homo* 11). Agamben presents the paradoxical guise of the animal-human, living-dying being, that is unable to comprehend itself. The recognition of bare life, knowing that knowing the animal and death is impossible, breaks cyclical anthropocentrism.

Cary Wolfe glosses Agamben's naked life in a slightly different, but complementary way when he explains how the animal connects with an unknowable, but palpable death. He writes that if humans *knew* death, then they would know the abyss—the empty interval between the human and animal—and, in turn, "the ethical relation to the other would be immediately foreclosed" ("Exposures," 21). Wolfe comments on the impossibility of knowing death: that it is in "just this impossibility that the possibility of justice resides: the (as it were) permanent call of the other in the face of which the subject always arrives 'too late'" ("Exposures," 22–3). In the Ojibwa origin myth, the dog has occupied the bare life, the threshold space between the subject and the other. Having experienced the world of the animals, it crosses the dark abyss and chooses to live alongside the humans. As a threshold figure that signals the chasm between humans and animals, the dog lets people know they will always arrive too late in the sense that they will be forever separated from the animal. Septimus Smith experienced this abyss, something too horrible to comprehend, but always close at hand. He is thankful that the dog does not continue to change into a human; heaven was "infinitely benignant." He will live another day, knowing mortality but not knowing it, living in death, but living.

With the notion of a separation between the animal and the human, the simultaneous ability and inability to know death in Agamben's hybrids, and the Ojibwe dog, let us turn to the dog in *Las meninas* by Diego de Velázquez and call it the "cynosure" of the painting. The cynosure—literally, "dog's tail"—refers to the Canis Maior or North Star and is a common term used in art criticism to signify the locus where all attention is focused. In the following pages, the dog will be the guide for the navigation and journey through *Las meninas* (Figure 4.1).[1]

One cannot deny that the dog had a long history as fidelity emblem. Cervantes put it well through the voice of Cipión in "The Dialogue of the Dogs:"

> People go on about our strong memories, our sense of gratitude, our great fidelity—so much so that artists sometimes use us as symbols of friendship. If you look, you'll notice that in those marble crypts where you see statues of the dead buried inside, whenever it's a husband and wife there's a figure of a dog at their feet. They mean this to show that their love and fidelity to each other know no bounds while they were alive. (24)
>
> Lo que yo he oído alabar y encarecer es nuestra mucha memoria, el agradecimiento y gran fidelidad nuestra; tanto, que nos suelen pintar por símbolo de la amistad; y así, habrás visto (si has mirado en ello) que en las sepulturas de alabastro, donde suelen estar las figuras de los que allí están enterrados, cuando son marido y mujer, ponen entre los dos, a los pies, una figura de perro, en señal que se guardaron en la vida amistad y fidelidad inviolable. (300)

Various critics read the dog in *Las meninas* as a fidelity emblem. J. A. Emmens reads *Las meninas* as a mirror for princes and argues that the dog represents the faithfulness of the ideal prince. Emily Umberger's also shares Emmens' interest in the dog as emblem of faithfulness, but sees "faithfulness" as not having a single reference (the ideal prince, or Philip IV), and as referring to various layers of meanings (faithfulness of the king to his wife, of the king to his painter, of the painter to his representation). Neither critic, however, extensively documents these assertions. Indeed, the dog as fidelity emblem in *Las Meninas* is suggestive, but ultimately does not have sustained support when read in the context of the painting itself, Velázquez's other works, and his source materials. Just as Cervantes recognizes the dogs' symbolic connection to fidelity but designs his dogs with another purpose in mind, so Velázquez was well aware of the artistic practice that connected dogs to fidelity, but he is more interested in exploring the tomb itself, rather than "fidelity," when he paints the brindled dog in *Las meninas*.

Velázquez was interested in the question of the representation of the animal ever since he painted the portrait of the stag that hangs today in El Prado. He also seemed interested in the animal by deciding to paint the portraits of Aesop and

[1] Critics do not consider the dog to be the painting's cynosure. Fred Licht states that the mirror is the cynosure (127) and Alban Forcione states that Margarita is the "central cynosure" (*Majesty* 6).

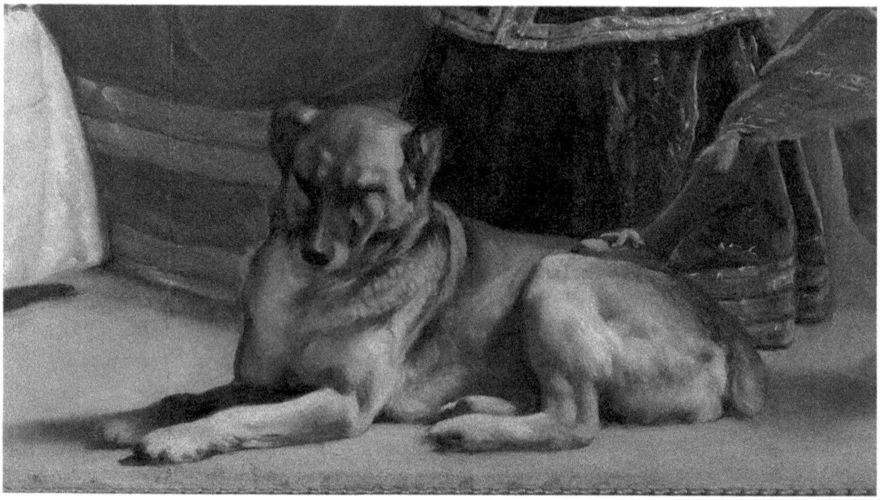

Fig. 4.1 Detail of dog in *Las meninas*, 1656, Madrid, Prado.

Mennipus, known as a Cynic or "little dog" and who satirized systemic philosophy.[2] In *Las meninas*, Velázquez does not just abandon the dog's symbolic connection to fidelity, but also abandons three other pictorial traditions: the dog as companion to the humanist scholar, to a noble figure, and to a dwarf. Instead, as I will argue, Velázquez associates the dog with approaching death, influenced by popular medieval sources, other paintings, and the emblem tradition of the cynocephalus. Velázquez's representation of the dog is akin to Agamben and Wolfe's description of the animal as a chiasmic living death. Every time that King Philip IV approached what appeared the living body of his favorite canine in the painting, he arrived too late. He could not touch a live animal, but only an artifice, a painted surface. Moreover, the dog, I suggest, warned the king that death approached in the sense that it told him of a simultaneous intimate knowledge of and incomprehensibility of death, especially with respect to his empire and family's future.

Early modern European artists—such as Albrecht Dürer, Abraham Hondius, Paulus Potter, Adriaen van der Velde—attempted to capture canines sleeping.[3] Connected to literary portrayals of the humanist and his dog (as retriever of ideas

[2] Like Velázquez, Cervantes was also interested in the Cynics. While the dogs carrying torches principally connected them with Dominican iconography, it also connected them with the Cynic philosophers.

[3] See Johns for illustrations. Another curious slumbering dog can be found with a set of sketches in the Goya Room of the Spanish National Library in Madrid. In 1906 the curator of the collections, Angel M. de Barcia, purchased the set of pencil sketches since he believed that one of the sketches, "El perro," was by Diego Velázquez and used as a workshop model for the dog in *Las meninas*. A twentieth-century Spanish historian of sketches would later write that "El perro" was not done by Velázquez, but a seventeenth-century Velázquez imitator (Pérez Sánchez).

such as in the case of *La Celestina* and Huarte), visual images of the slumbering dog in the sixteenth century were also emblems of the humanist mission and intellectual labor. Dogs at the feet of the humanist scholar symbolized the scholar's faithfulness to his enterprise (Höltgen). While the humanist scholar works, a curled-up or sleeping dog rests at his feet. A common sixteenth-century representation depicts a slumbering dog (not just a lion) with St. Jerome at work on intense philological exercise. Posthumous portraits of Petrarch (who not only wrote important love poetry, but also was one of the first to translate and revive ancient Greek) show him at work in his study with a lying dog. Although Petrarch worked in the fourteenth century, the fame of Petrarch's love for his canine companion Zabot was widely known in the renaissance and the lying dog in his portraits celebrates the humanist work of finding and breathing life into the classics (Reutersward 54–5; Schiesari, *Beasts* 32–43). Another visual example of the dog as companion to the humanist can be found in the frontispiece of *Four Books of Love* (1502), a Latin work by Conrad Celtis (Reutersward 54).

Justus Lipsius not only writes a letter around 1600 to his students in which he expresses friendship and affection for his dogs, but he also explains to his students that the dog is the ideal symbol for the humanist at work because the humanist studies days and nights on end like the watchful dog and must stick to his studies with the physical stamina, intelligence and loyalty of a dog (Papy 170). In pictorial traditions, Lipsius was often portrayed with dogs and in one image—a posthumous allegorical glorification of the scholar—a dog lies beneath his entombed body (Papy 183).

Tycho Brahe carefully recorded the position of the stars in *Instruments for the Restoration of Astronomy* (1598). One illustration of his book is that of a mural quadrant that allowed the observer to measure the altitude at which celestial bodies cross the meridian (Estrin). The illustration includes a portrait of Brahe himself with the iconic dog resting at his feet (Figure 4.2). Explaining the symbolism of the humanist dog, Tycho Brahe writes that the dog lying at his feet is a symbol of the pursuit of intelligence and loyalty to the scholarly endeavor (Reutersward 55–6).

Affirming its important role as humanist icon, painters in renaissance Italy such as Tintoretto illustrated positive conceptions of the canine character and faculties such as superior perception and exemplary rationality (Cohen 280). Philip II, Philip IV's grandfather, admired the Biblical figure of Solomon. Solomon is depicted in a great fresco at the center of El Escorial's library and, in his painting *The Visit of the Queen of Sheba to Solomon* (1559) at the Cathedral of Ghent in Belgium, Lucas de Heere uses a portrait of Philip II's face for the face of the figure of Solomon. Scholars comment that Philip II not only admired Solomon for his intelligence, but that he also had a dog that he named Solomon (Cuadra Blanco). The name of Philip II's dog, Solomon, shows that Philip II projects his high appreciation of Solomon onto his dog. In Sint Janskerk de Gouda, a Gothic church in the Netherlands, a stained glass window depicts Philip II and Mary Tudor seated with a white dog (1557–59) (Cuadra Blanco). The dog in the stained glass invokes the sixteenth-century humanist connection between the dog and study since the dog, whose name is possibly Solomon, has his muzzle in basket of letters and papers.

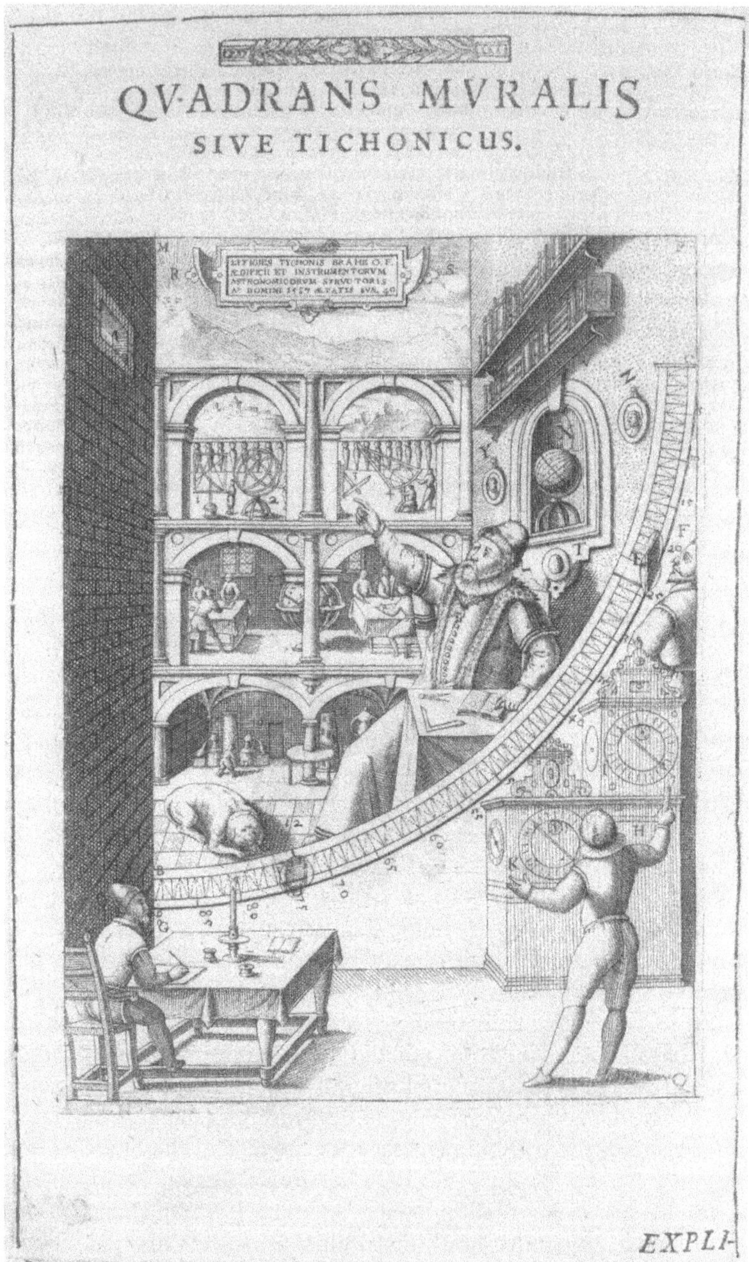

Fig. 4.2 Plate illustration from Tycho Brahe's *Astronomiae instauratae mechanica*, 1602. Courtesy of the Smithsonian Institution Libraries, Washington, D.C.

In *Las meninas*, the slumbering dog in the company of the pensive self-portrait of Velázquez evokes the humanist scholar tradition. Leo Steinberg suggests that on one level the dog's body transfers the humanist's enterprise to that of the artist's. Like the humanist, the artist in *Las meninas* pauses, imagines, and creates a portrait. In this sense, according to Steinberg, the dog is a symbol of the sacred silence (or *apathes*) that accompanies the mental labor of the humanist (50). But Steinberg does not comment on the ways in which Velázquez diverges from the sixteenth-century representation of the artist and dog, adding new elements to the humanist-dog tradition. The painting of the dog in *Las meninas* radically reconfigures the humanist-dog icon tradition. For instance, Velázquez depicts something never seen in the humanist-dog iconic tradition: a moment of physical human contact with the animal.

Erica Fudge writes that Derrida tries to "uncover a different kind of human and a different kind of philosophy from the one that has placed the dog at its feet" ("The Dog," 38). In what follows, I argue that Velázquez quite literally represents the dog in a different way than the humanist pictorial tradition that "placed the dog at its feet." Velázquez does not depict the artist *alone* with the animal. He introduces a cast of characters that occupy the canvas along with the artist, such as Nicolás Pertusato, the Italian court jester, often called a midget or dwarf. Unlike the image of Velázquez, who is separated from the dog, Pertusato not only touches the animal, but he gives it a nudge with the toe or side of the foot.[4] As opposed to the humanist who finds inspiration in the image of a sleeping dog at a distance, Nicolás pokes the animal with his foot as though to wake it up, a movement that signals that Velázquez paints an image that both remembers and might "wake up" from the tradition of painting the humanist in his study.

While *Las meninas* seems to portray a dog shaking off the slumber of the sixteenth-century icon of the humanist, it most directly evokes and abandons the popular sixteenth century artistic genre of the dog-noble portrait. Philip IV would have been aware of the pictorial tradition that celebrated the relationship between nobility and dogs. The Italian master Titian painted a small dog with *Federico II Gonzaga, Duke of Mantua* (1529–30) which hung in the South Gallery of the Royal palace (today it is in the Prado). The dog suggests the faithfulness of its owner. Other paintings that included dogs highlighted not just the master's faithfulness, but also masculinity and power of the Spanish royal family. In the Escorial, Velázquez would have seen the most famous royal-dog portrait of a Habsburg. In Titian's portrait of Charles V, the emperor stands near a hunting dog (Figure 4.3). The dog as royal companion symbolically heightened the owner's

---

4 *Las Meninas* calls attention to the dog's body in the present and past. A human foot touches it. The body of this Spanish mastiff was violently altered in the past because a human hand cropped its tail and ears. Herrera writes that the best dogs are the ones that put up with the most suffering: "they say that the sign of a good dog is when you come up to it and grab it by the ear; if it does not growl, it is the best" ("... dicen que es señal de muy buen perro, que quando chico le asgan de la oreja con la mano, y le alcen, y el que mas lo sufriere sin gañir es mejor" [my trans.; 348]).

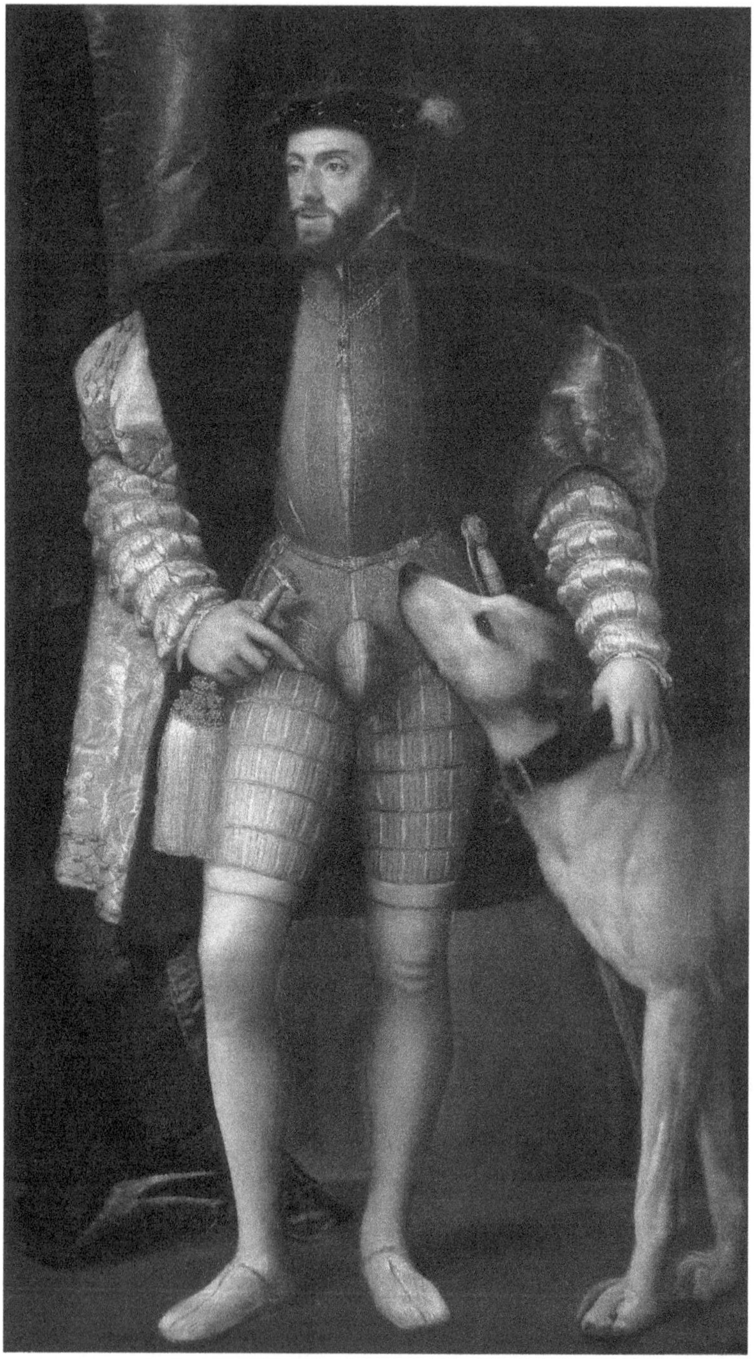

Fig. 4.3 Titian, *Charles V*, 1533, Madrid, Prado.

humanity, as it leant credibility and elevated status by inserting him into a tradition of noble dog portraits.

Paralleling and complementing the iconography of the slumbering dog with the humanist, the image of the dog in Titian's portrait of *Charles V* connotes both masculinity and faithfulness. In the portrait tradition that borrowed from the "mirror of princes" tradition, the exemplary monarch poses alongside a dog that emblematizes the king's faithfulness as ideal prince. In Spain the portrait of the large dog with the Habsburg in *Charles V* did not just make its owner more powerful, masculine, and faithful in an abstract sense. The large dog was often equated with the lion. Many cultures conflate the lion and the dog as evidenced in the lion-looking Chinese Dogs of Fo or the Japanese *shishi*. In the early Spanish zoological mindset, the word for a canine puppy was the same as a lion cub. As Covarrubias notes, *cachorro* (puppy) is only reserved for the progeny of big dogs and it is the same word used, by association, for lion cubs (258). The sixteenth-century writer Herrera provides a series of Spanish dog names that should be given to one's Spanish mastiff. He writes that dogs understand much of what is told to them and that they come when they are called by their name (347). They should be given names when they are young so they know their names well and the name should be short, no longer than two syllables since the shorter name will help them heed their master more quickly. Among possible names, Herrera suggests "León" is a good option (347).

The Spanish mastiff was given a noble genealogy by Herrera. Because of its legendary ancestry that connected it with symbolically powerful animals like the wolf, tiger and lion, the Spanish mastiff came to be thought of as an aristocratic and regal animal. Herrera explains that the Spanish mastiff's pedigree, its *casta*, could include other animals such as the lion. Herrera writes that the Spanish mastiff is part wolf and, borrowing from classical sources such as Pliny, he adds that some specimens were part lion or tiger:

> Oftentimes wolves will mate with Spanish mastiff bitches that have birthed or are in heat. Moreover, in some lands like India, lions or tigers will mate with them and the *casta* that results from these mixes is quite good and strong … (my trans.)
>
> Las perras estando paridas o calientes se toman muchas veces de lobos, y aun en muchas partes, como en la India, se toman también de leones y tigres, y de la casta que sale de aquellas mezclas, y ayuntamientos salen muy buenos, y recios perros … (350–51)

The mixed breed origins of the dog lent it a monstrous quality, similar to a mythological animal, making it awe-inspiring.

Sometimes the mastiff's aristocratic reputation was connected with classical literature, making it a symbol of both majesty and the humanist enterprise. Through the use of a simile in "The Glass Graduate" (El licenciado Vidriera), an exemplary novel included in the same collection as "The Dialogue of the Dogs," Miguel de Cervantes affirms the Spanish mastiff's reputation as an old stock dog purebred, a type that was connected with gravitas. He writes that contemporary poets are like little barking dogs while the classic poets are like old and serious great Spanish mastiffs. In a hilarious scene in which the protagonist Vidriera reads a sonnet to his audience, he contrasts his profound mastiff-like sonnet with contemporary doggerel: "And what shall I say of the thefts committed by these cubs and whelps of modern pretense from the grave and ancient great Spanish mastiffs of the poetic art" ("¿Qué diré del ladrar que hacen los cachorros y modernos a los mastinazos antiguos y graves?" [80; 59]).[5]

Across Europe, when the nobleman posed with his dog, especially a large hunting dog, his body was humanized in the sense that it bestowed upon the owner a sense of dignity because the hunt was considered a dignifying and ennobling practice.[6] In the noble-dog portrait, the image of a large dog also infused its companion with a sense of physical and imperial power, especially in the Spanish context since it transferred to its owner the spirit of the American conquest. Although the dog was not the primary weapon in the Spanish success in the Americas, sixteenth-century sources consistently signal the dog as embodying key attributes of the character of the ideal conquest hero. Epic poetry such as Juan de Castellanos' *Elegies of the Illustrious Men of the Indies* (*Elegías de varones ilustres de indias*) praised large dogs as winning key battles. These dogs had many names, including Leoncico, the diminutive form of León.[7] As mentioned in Chapter 1, dogs were also used to keep colonial order and executed punishments.

Many sixteenth-century histories of the Americas describe large Spanish dogs that had a special ability to reason in contrast to the irrational natives. Oviedo writes one anecdote about non-cannibalistic Spanish dogs that were more human

5 I have modified the translation. The English translation erroneously has "masters" instead of "Spanish mastiffs."

6 See Mateos' *Origin and Dignity of the Hunt* (*Origen y dignidad de la caza*) for a hunting treatise from the period that uses the word "Dignity" to describe big game hunting with dogs.

7 Aside from Leoncico, other famous dogs in the Americas recorded in the colonial Spanish texts include: Bruto (Brute); Turquillo (Little Turk); Amigo (Friend); Menalao (Menaleus); Amadís (Amadis of Gaul); Becerillo (Little Calf); and Lautaro. For an account of the Bruto, see Alves 156. For an account of Turquillo, Amigo, Menalao and Amadís, see the epic poem written by Castellanos (Miramón 124–25). Castellanos mentions that in the Nuevo Reino de Granada (modern-day Colombia) the Spanish soldiers Juan Meléndez, Juan de Vargas, and Baltasar Núñez were accompanied, respectively, by dogs named Turquillo, Amigo and Menalao. Castellanos also writes of the death of Amadís, who suffered for a grueling 24 hours after injury by arrows. Many sources describe the father, Becerrillo, and son, Leoncico, the most famous of the conquest canines. Becerrillo is known for its feats of wars in the Caribbean and is often attributed as belonging to Ponce de León (Miramón 122). Guaman Poma de Ayala also has a dog called Amigo and one named Lautaro, both of whom are found in his self-portrait walking to Lima at his side with his son. Did Guaman Poma name Lautaro in celebration of the famed Mapuche Araucanian leader (1534–57), described in Alonso de Ercilla y Zúñiga's epic (1569)? Interestingly, with respect to the animal and names, "Lautaro" is the Spanish pronunciation of two Mapuche words meaning "swift caracara" (a falcon-like bird of prey).

and civilized than the savage natives since the Spanish dogs would not touch dog jerky, a common staple among the natives.[8] In Francisco Cervantes y Salazar's *The Chronicle of New Spain* (*Crónica de la Nueva España*) (ca. 1560) the Spaniard soldier Peñalosa owns a dog that he would die for. Cervantes y Salazar writes that the dog would face 10,000 Indians and it would not hesitate in launching itself at them, clutching their throats in its jaws. When the Aztec leader Cazonci captured

[8] Gonzalo Fernández Oviedo describes Spanish dogs as trained warrior dogs and tells the story of how in Nicaragua the natives typically ate dog meat. In one province the people prepared a special dried dog jerky whose name *xulo* (a semantic cousin of the breed Xolo and identity category Cholo). Oviedo writes that the Spanish dogs that happen to come close to the native dog jerky will have nothing to do with it. Oviedo explains that he tried the experiment himself: he took seven or eight Spanish dogs and put the dog meat in front of them. The dogs were repelled by the sight: they only smelled the meat, but did not touch it or eat it and did not even want to see it. Oviedo concludes the anecdote by contrasting the incident of the Spanish dog's aversion to eating dog meat with human cannibalism: "The Indians do not show the same courtesy or customs with human flesh since they eat one another" ("La cual cortesía y comedimiento no usan aquellos indios con la carne human, pues se comen unos a otros" [my trans.; 4: 426]). But this canine–human cannibalism connection is not true throughout his chronicle. Oviedo provides the earliest example of dog-eating by Spaniards: "In that time the Christians were so famished that they ate all the *gozque* dogs that were on the island, both the mute ones and the ones they had brought from Spain" ("En este tiempo de tanta necesidad se comieron los cristianos cuantos perros gozques había en esta isla, los cuales eran mudos que no ladraban, se comieron también los que de España habían traído" [my trans.; qtd. in Alvárez López 413]). Other chronicles report Spaniards eating dogs to survive. An account of the De Soto expedition indicates the Spanish consumed dogs while the natives did not: "The Indians at Guaxulle made the Governor service of three hundred dogs, for they observed that the Christians liked them and sought to eat them, but they were not eaten among the Indians" (qtd. in Werner 4). When Cabeza de Vaca arrived to southwest Texas in the late 1520s, he mentions that he traded pelts for dogs with the natives so that his party could have enough nutrition to continue their journey. He also mentions that he stole a dog and ate it from a people that would have never given thought to eating a dog even though they were barely surviving from a hunter-gatherer lifestyle. Also, accounts from the Americas sometimes indicate the same disdain for dog eating that Spanish chronicles do. Aside from Spanish texts, some Quechua texts indicate that native peoples disdained other native cultures that ate dogs, a practice that was also associated with human sacrifice. The Huarochí manuscript (a colonial era Quechua text compiled by Francisco de Ávila at the end of the sixteenth century) disparages cannibalism and connects it to dog-eating: "Because he fed on people, let him now eat dogs; and let the Huanca people worship him" (*Huarochí* 110). The Inka peoples prohibited the practice of sacrifice and eating dogs, but their enemies, the Huanca peoples continued to propitiate a fire monster with dog sacrifices. Some chronicle accounts record that the Huanca continued to worship a dog god and their most important feast was a dog banquet. Quechua-Hispanic chroniclers such as Felipe Guaman Poma de Ayala, who, as mentioned, portrays himself as traveling with a dog who he calls "friend," understood the Huanca's dog-eating in cannibalistic terms: it was a substitute for having previously eaten humans. Like Muñoz Camargo, Guaman Poma writes that the practice is prohibited by law: "although dog-eating is now considered against the law, some of these Huanca people still risk punishment by carrying on the old custom" (qtd. in M. Schwartz 74).

and sacrificed the dog, Cervantes y Salazar concludes that Peñalosa's dog's death is "as sad as the case of unfortunate, blameless human deaths; (the dog sacrifice) was, verily, a horrible thing that defies all natural reason upon telling it, and even more so upon having seen and experienced it" ("tan tristemente como en las tristes muertes de los que no eran en culpa dellas solían hacer, cosa, cierto, espantosa y que la razón natural rehuye contarla, cuanto más verla y hacerla" [my trans.; Cervantes y Salazar 271]).[9]

The image of the large dog when paired with a royal figure lent its owner the symbolic potency of the Spanish imperial conquest as well as the intelligence behind the success of the conquest. With respect to its connection with royal character, as mentioned in the previous chapter, an anonymous poet gives the name León to the protagonist of his soldier dog in the poem "El Alano." Little Lion (Leoncico) was also the name of the most famous soldier dog in the Americas. Little Lion was known as a *perro sabio* (wise dog) and is recorded to be the conquistador Núñez de Balboa's favorite. Under the heading, "Concerning the Dog that Earned a Crossbowman's Salary" ("De un perro que ganaba la misma ración que un ballestero" [3: 444]) in *The General History of the Indies* (*Historia general y natural de las indias*), Oviedo writes that Leoncico "earned the same as or, at times, double the amount of successful soldier" ("ganaba, asimismo una parte y a veces dos, como los buenos hombres de guerra" [my trans.; 3: 103]). Attesting to the popularity of León in sixteenth-century Spain, the humanist dialogue *The Garden of Curious Flowers,* Antonio de Torquemada describes Little Lion as having an uncanny ability to communicate with people.[10]

---

[9] I should note that Spanish dogs in the Americas were not always portrayed as soldier or punishment animals as has been perpetuated by Todorov or in popular histories such as *Dogs of Conquest* in which John Grier Varner and Jeannette Johnson Varner provide an extended study of the history of Spanish dogs in the Americas as soldiers and punishers. Indeed Garcilaso Inca de la Vega describes how one colonist treated his Spanish mastiff like a pet. In *The Royal Commentaries of the Incas* he writes that colonists had no need to bring Spanish mastiffs to Peru since there were no wolves or other predatory threats to livestock. Nonetheless, one colonial dog owner makes it his job to be a substitute nurse for his *mastín* puppy: "Even though there was no reason, they brought them any way; one Spaniard brought a month-and-half-old Spanish mastiff puppy the 120 leagues of treacherous trail from Cuzco to Reyes. He carried it in his saddle bag that was hanging on the front of his saddle tree; every day the puppy gave him new work since he had to find it milk" (...no haber para qué, no más de por el bien parecer, trajo un español desde el Cozco hasta los Reyes, que son ciento y veinte leguas de camino asperísimo, un cachorrillo mastín que apenas tenía mes y medio. Llevábalo metido en una alforja que iba colgada en el arzón delantero; y a cada jornada tenía nuevo trabajo, buscando leche que comiese el perrillo [my trans.; 361]).

[10] The sixteenth-century dialogue describes "Little Lion" and his uncanny ability to communicate with people, conflating him with stories about his father Becerrillo: "How can one judge that dog called Leoncico who travelled with a soldier when Columbus began discovering the West Indies? He fought in all sorts of battles and the Indians confessed to be more afraid of that dog than twenty Christians. Moreover, if some Indian of those that the

Aside from its symbolic connection to the imperial conquest, the large dog was a symbol of Spain's economic success. The carefully-crafted design of the *mastín* (Spanish mastiff) played a crucial role in the economy that made the Spanish empire possible.[11] The production of wool and textiles gave Spain the necessary wealth to launch its imperial enterprise. Carla Rahn and William D. Phillips write: "Spanish herding was not just a source of supplemental income, it was a crucial source of developmental capital for the economy as a whole" (xii). In the late medieval period, the Spanish textile industry exploded due to the high quality of the Spanish merino sheep variety and the sale and production of its fine wool. In the sixteenth century, Spain had over three million sheep and exported them across Europe and to the Americas; indeed one of the primary economic dreams of the empire was to successfully implant Spain's sheep Mesta economy on American soil (Melville).[12] The Spanish mastiff contributed to the maintenance and production

Christians had captured got free and ran away, then they would simply report the incident to Leoncico as if they were talking to another man..." ("¿Qué se puede juzgar de aquel perro llamado Leoncico, que pasó con un soldado cuan Colón comenzó el descubrimiento de las islas occidentales, el cual peleaba de tal manera en las batallas, que confesaban los indios tener muy mayor temor del perro que de veinte cristianos? Y lo mas era, si algún indio de los que habían prendido se soltaba y huía, no hacían mas que decirlo a Leoncico como si hablaran con otro hombre...?" [my trans.; 345]).

11 Herrera's complete description reads as follows: "The (best Spanish mastiff) should have a big head so that it seems to be a third of the animal's body and that looks like the face of a man's. It should have a wide and open mouth, full lips that hang from the mouth, and big, floppy ears. Its eyes should be black and not light blue. The eyes should also be shiny and alive like sparkling stars. The Spanish mastiff should have a great and terrifying bark and a wide breast and back. Its neck should be thick and short, that is, it should have a very short neck. Its body should be short, rectangular and not long. Its legs thick and furry. Its toes should be long and separated and they should extend through its paws. Its tail should be short and thick since it is a sign of its strength" ("Han de ser de grande cabeza, tanto que parezca tener, o ser un tercio del cuerpo, la cara que parezca de hombre, muy grande boca, y muy ancha, y muy abierta, los bezos grandes que cuelguen de la boca, las orejas muy grandes, y caídas: los ojos muy relucientes, vivos que parezca que centellean, prietos, y no zarcos, de muy grande ladrido y espantoso, de muy ancho pecho, y espalda, el cuello gordo, y muy corto, el cuerpo corto, cuadrado, y no luengo, los brazos gordos, y bien vellosos, los dedos largos, y bien partidos, y que asienten muy bien todo el pie y mano, y la cola corta, y gorda, es señal de más fuerza" [my trans.; 347–8]). Herrera also highlights the breed's distinctive dew claw when he writes that "there are some that they call 'claw footed' since they have an extra dew claw and those are stronger than the others" ("hay unos que llaman pesuñados que tienen un dedo atrás, aquellos son muy más recios que los otros" [my trans.; 348]). Contemporary descriptions of the breed in dog manuals still highlight the breed's distinctive double dew claws on the hind feet (Benito Ruiz and Malo Acrudo).

12 Imitating the Spanish imperial economic model, George III of England would send Spanish merino sheep to North America and Australia.

of sheep and, because of its importance, was often treated in human terms.[13] Since approximately five Spanish mastiffs protected every 1000 head of sheep during the peak wool years, 1477–1563, one scholar has concluded that there were over 10,000 of these dogs guarding sheep in Spain and its colonies (Delgado Díaz).[14]

Images of the large dog posing next to the Spanish monarch would have signaled the monarch's strength in general and, when that dog was a *mastín*, the dog would have suggested the economic success of the monarch. Moreover, the symbolic charge of the large dog alongside a royal figure evoked the most significant pendant favored by the Hapsburgs of the period. Artists often depict Spanish monarchs wearing this pendant known as the Badge of the Order of the Golden Fleece. [15] A portrait of the Spanish king who posed with a large dog and specifically a *mastín* evokes the portrait tradition of the Spanish monarch wearing the badge of the Order of the Golden Fleece. Although connected to the mythological Jason and the Argonauts and originally from Northern Europe, the

---

[13] Just as the Spanish war dog was treated in human and humanizing terms—earning a soldier's salary and communicating as a person—so the Spanish mastiff was treated in terms of human dignity. Sheep dogs received the same food allotment as a human shepherd; two pounds of bread a day and fines of five sheep or more a day were exacted from those who injured a sheep dog (Klein 25; Alves 84). Even the physical design of the animal was conceived in such a way that its face should appear like a human's. As Herrera writes: "The (best Spanish mastiff) should have a big head so that it seems to be a third of the animal's body and that looks like the face of a man" ("Han de ser de grande cabeza, tanto que parezca tener, o ser un tercio del cuerpo, la cara que parezca de hombre" [my trans.; 347-8]).

[14] Since it played a fundamental role in maximizing wool production, breeders sought techniques to produce the finest possible Spanish mastiff to protect the flock. Herrera writes that it is also good idea for the shepherd to have a *galgo* or other type of rabbit-hunting dog since they can run faster than the Spanish mastiff and reach a wolf that threatens the flock much more quickly. Herrera gives detailed advice about how the shepherd should take care of his Spanish mastiff such as when to feed, how to breed, what to do about rabies and other diseases, how to avoid flies in the summer time, and how to crop its tail and ears to make it more difficult to attack for the wolves that attack the flocks. Herrera provides another reason for cropping the tail when he writes that if it is short and think then the animal will be an even stronger specimen. In Spain the herd did not follow the dog, but followed the bell of the tame wethers that accompanied the human herder (Rahn and Philips 105). The Spanish mastiff was bred as an intensely territorial guard dog. Descendants of the Spanish mastiff were still described as protecting sheep in the Southwest in the nineteenth-century United States. Descriptions of these immigrant Spanish mastiffs characterize them as having a supremely gentle demeanor with the sheep, but, in contrast, powerfully ferocious when protecting the flock. Accounts indicate that the Spanish mastiff easily guarded the sheep from wolves, bears and marauding dogs (Derr 44).

[15] Various artists portrayed Spanish Habsburg Kings wearing the badge of the Order of the Golden Fleece. Philip IV is portrayed throughout his life wearing the badge of the Order of the Golden Fleece including Gaspar de Crayer's *Philip IV in Parade Armor*. Diego de Velázquez painted Philip IV wearing the badge of the Golden Fleece at least three times. One of these portraits was completed the same year as *Las Meninas*.

badge of the Golden Fleece, a small medallion in the shape of a hanging sheep skin, alluded to the success story of the imperial Spanish economy, the fleece of Spain's prized merino sheep. The connection between the badge and Spanish sheep was more explicit in other contexts. The conquest of America, for instance, was conceived of as a realization of Jason's quest. Many chronicles describe the Spanish conquerors in the Americas as the *true* heroes of the Golden Fleece, not Jason who, incidentally, was accompanied by a famed dog (Gerbi 270). When a kingly figure poses next to a large dog, especially when the figure is Spanish royalty, the dog signals Spain's military might, its economic success, and other iconographic emblems and portraits of imperial power such as King Philip IV wearing the Badge of the Golden Fleece.

But the dog in *Las meninas,* while invoking kingly power, only does so to signal its demise. With the dog in *Las meninas*, just as it calls forth the humanist enterprise and unsettles it, so it also reminds the king of his faltering economic and military enterprise. By 1656, when Velázquez painted the Spanish mastiff in *Las meninas*, the market for Spanish wool had radically fallen and subsequently the Spanish mastiff as shepherd dog found itself without a job in the sense that less and less sheep motored the Spanish wool economy. In his 1961 play, also entitled *Las meninas,* Antonio Buero Vallejo names the Spanish mastiff in the painting "León." Although no archival document mentions the name of the dog, Buero's name makes historic sense. When one character looks at the León, he alludes to the fact that the Spanish lion has now become only a dog: "He's tired. He reminds you of a lion, but the Spanish lion is now nothing more than a dog" ("Está cansado. Recuerda a un león, pero el león español ya no es más que un perro" [my trans.; 101]).

In contrast to the many portraits of King Philip IV wearing the Badge of the Golden Fleece, Velázquez may be using the mysterious canvas in *Las meninas*—whose reverse side faces the spectator—to *suggest* a royal portrait. In his evocative study, Keith Broadfoot writes that he imagines the portrait of King Philip IV as the hidden image of the canvas. He then makes an observation with respect to the death of the king:

> If one is drawn to examine the evidence of this side of the canvas, one notices what appears to be an accidental pattern of stains; and if one remains captivated by these stains for a moment, one cannot help noticing the dark patches of a pair eyes and the vertical slash of a nose—in short, a death's head. Whose death is figured here? (231)

Although Broadfoot's suggestive comments about what is painted on the hidden side of the canvas are largely conjecture, Velázquez does paint a somewhat visible monarch: the portrait of the king and queen appears in the mirror in the background. The king and queen may have been literally posing next to the dog when Velázquez painted *Las meninas*, but the painting itself does not represent the king next to the dog. Velazquez interestingly *suggests* the tradition of the noble-dog portrait and the monarch's power. In fact, Bo Valne has studied the position of the figures in the painting and writes that king and the queen had to have been posing right where the dog is lying because, from his position just behind the canvas, the painter would have positioned himself in a place where he would just have to turn his head slightly to see them (167). In this sense, the dog in *Las meninas* makes a reference to the royal-dog portrait tradition and, also, indirectly, the power, faithfulness (as J. A. Emmens suggests), the economic power of the empire, and the powerful, special bloodline of the king.

Just as Velázquez *implies* that the king is next to the dog, he partially removes him from the painting. The King and Queen are blurry images in the mirror at the back of the painting. In this sense, the dog-king portrait in *Las meninas* is radically different from Titian's portrait of Charles V standing next to his dog. *Las meninas* does not highlight his master's faithfulness, power and virility as it had done in the Titian portrait in which Charles V's codpiece stands out near the dog's outstretched head. Velázquez suggests, I would argue, that, in contrast to *Charles V*, the king in *Las meninas* is emasculated in the sense that the portrait no longer depicts the king posing with his dog. Further, the separation of the king from his dog in *Las meninas* is a sign of the death and dying of his livelihood and power: his vast empire is losing ground to England and the Netherlands; his economy is in shambles; and his royal bloodline is losing its reputation.

A survey of some of Velázquez's other paintings reveals the ways in which the portrait of the dog in *Las meninas* connects to the approaching death of Philip's progeny. Velázquez explored versions of the noble-dog genre throughout his life. He painted a royal figure with a dog, *Cardinal Infante Don Fernando as Hunter* (c. 1632–33), some twenty years before *Las meninas.* The painting, which prominently shows a *galgo,* was destined for the collection of royal hunter portraits for the decoration of the Hunting Sala at the Torre de la Parada in the royal hunting grounds of the Pardo. In 1899 Jacinto Octavio Picón wrote a biography and analysis of Diego Velázquez's works, saying of the *galgo* at Philip IV's brother side: "Just as they say that in well-wrought portraits the person is speaking, so this dog is not barking because he does not feel like it" ("Así como se dice de las personas bien retratadas que está hablando, pudiera decirse de este can, que no ladra porque no quiere" [my trans.; 76]).[16] The dog in *Cardinal Infante Don Fernando as Hunter*, while suggesting faithfulness and masculinity, evokes the separation between the king and his dog that will occur in *Las meninas.* Philip IV's younger brother stands next to the faithful greyhound holding his angled firearm, suggesting some of the grandeur and phallic potency of the Titian painting, but the wonderful, grey Greco-like cloud cover ushers in an ominous augur.

In *Philip IV as Hunter* (1633–36) (Figure 4.4), Velázquez paints the king with another Spanish mastiff. But *Philip IV as Hunter* connotes a sense of dejection and

[16] In her study of dogs and material culture, Adrienne Martin calls attention to magnificent collar that the dog wears in *Cardinal Infante Don Fernando as Hunter*. She notes the different collar worn by the dog in *Baltasar Carlos as Hunter*. She compares this decorative collar to the brass one worn by Berganza when he works for the constable. Martin also notes on the importance of special collar used protect the dog from animal attack that Berganza wears when he works as a sheep dog.

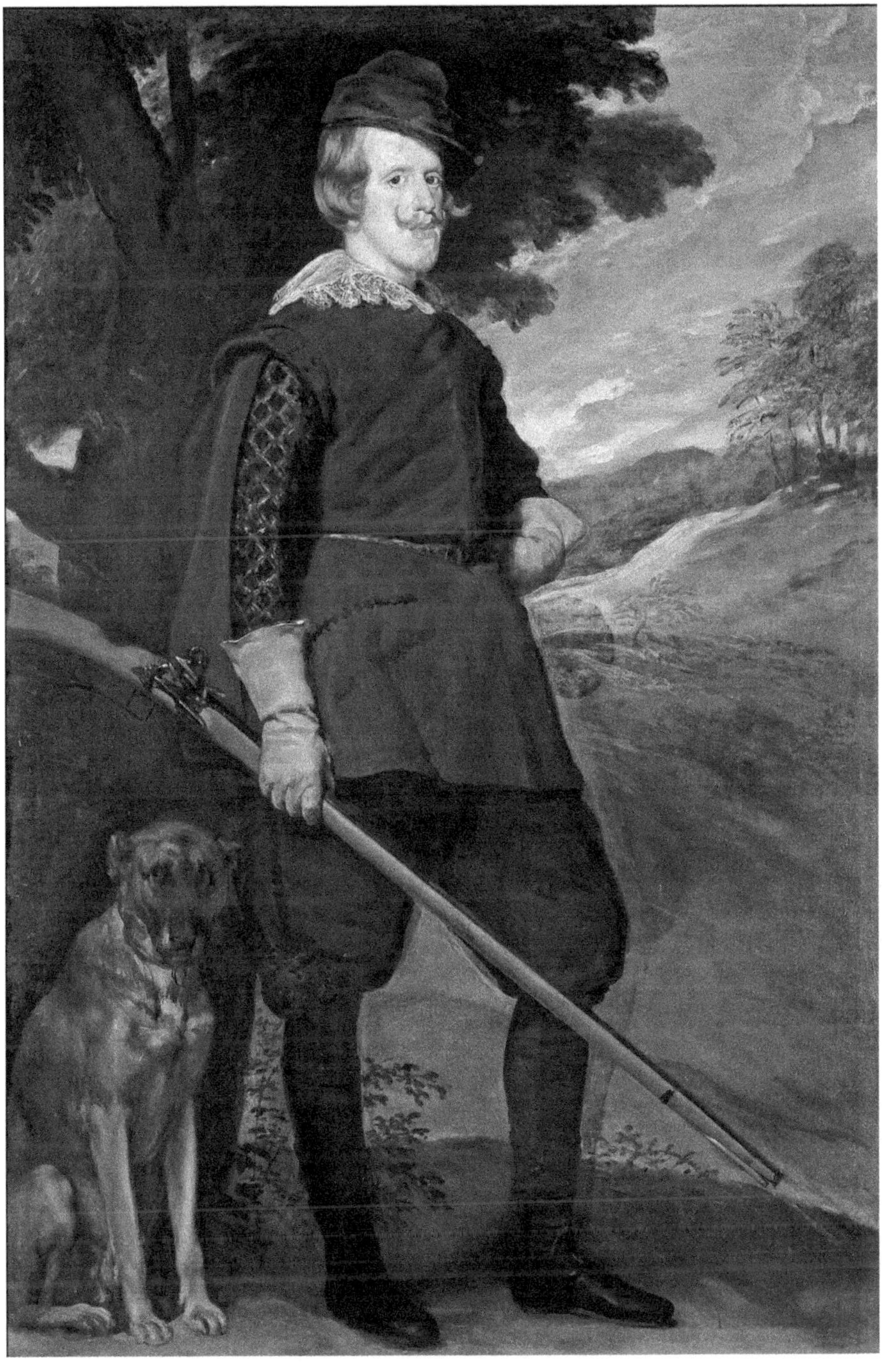

Fig. 4.4 Diego Velázquez, *Philip IV as Hunter*, c. 1636, Madrid, Prado.

failure by the lowered musket and the moping expression on the Spanish mastiff. The dog is not a greyhound-like hunting dog as in *Cardinal Infante Don Fernando as Hunter*. Since it is not a hunting dog, but a guard dog, the title of the painting misleadingly suggests that the Spanish mastiff is the king's hunting animal. Art historian Thomas Glen makes some fine comments on the dog in *Philip IV as Hunter,* but erroneously leads one to believe that the Spanish mastiff is a hunting dog. Spanish mastiffs served as guard dogs in the royal court and, as mentioned, were pastoral animals by training.

One may contrast the dog in *Philip IV as Hunter* with the large dogs alongside their masters in Titian's *Diana and Acteon* and *Charles V*. In each of these Titian paintings, the heads of the dogs are erect and suggest phallic potency. By including a Spanish mastiff in *Philip IV as Hunter*, Velázquez introduces the theme of the emasculated Habsburg which he explores even more openly in *Las meninas*. Philip IV, a Habsburg, like the mastiff, had a noble, majestic reputation (in the painting Philip IV is regally outfitted in hunting gear), but he did not "hunt." Philip IV enjoyed pomp, but did not exercise imperial leadership; he could be found at plays in the Zarzuela while his chief minister, the Count Duke of Olivares, executed policy (Beusterien, *An Eye* 54).

In these two earlier portraits, *Cardinal Infante Don Fernando as Hunter* and *Philip IV as Hunter*, Velázquez began to explore new meanings for the dog that departed from its symbolic role as the animal that humanized the figures of the humanist, the nobleman and the king. In a similar break from portraiture tradition, just as Velázquez rejects making the dog a royal emblem in *Las meninas,* he also does not use the dog to degrade the dwarf. He does not represent the animal to signal the dwarf as sub-human.[17] Velázquez had studied dwarfs as dignified subjects with the stateliness of royalty in portraits such as *Sebastián de Morra* (c. 1643–44) and *Diego de Acedo* (also known as *El Primo*) (c. 1633–34), capturing the subject's individual personality and piercing gaze of intelligence and intensity. Diego de Acedo is a humanist figure who sits with his hand marking the pages of a huge tome with a pen in front of him surrounded by other books. Neither of these two dwarf portraits shows a dog.

Moreover, a marginal anthropocentric portraiture tradition existed in which the dog accompanied the dwarf. Sometimes the portrait (such as Juan Pantoja de la Cruz's depiction of an Irish greyhound, named Baylan, with one of Philip III's dwarfs and the 1560 Netherlander Antonis Mor's *Estanislao*) highlighted the dog and dwarf as royal gifts, animalizing both as animal plaything in the court. In other works such as Edward Tiberio de Tito's *A Dwarf with Medici Dogs in Boboli Gardens* (1620–25) and Jan van Kessel the Younger's *Two Dwarfs of Philip IV* (ca. 1680), the small size of the dog imitated the small size of the men who accompany them. In *Las meninas,* Nicolás does not pose with the dog, but touches and

---

[17] Nonetheless, in other paintings, Velázquez does not necessarily treat the dwarf with respect. He for instance infantilizes the dwarf in *Baltasar Carlos and Dwarf* who holds a rattle in contrast to the young prince, holding a scepter.

interacts with the dog. Maribárbola, Nicolás, the artist, and Margarita, her maids (*las meninas*) and others all stand in the same room as the dog. Velázquez neither uses the dog to humanize and heighten the reputation of the royal sitter, nor, as in the case of the dwarf-dog genre, does he uses the dog to dehumanize small adults like Maribárbola and Nicolás.[18]

Velázquez had a special interest in the noble-dog portrait, especially when the noble personage was a child. *Baltasar Carlos as Hunter* (1635–36) fits into the traditional child noble-dog portraiture genre. Prince Baltasar is accompanied by hunting dogs, including a slumbering pointer and *podencos.*[19] The closed eyes of the lounging Old Spanish pointer contrast the alert expression of another *podenco* that is nearly truncated out of the right side of the painting. Only the *podenco*'s face and front legs are revealed and a workshop copy of this painting reveals the entire body of an additional dog, a picture of another *podenco* (Brown, *Velázquez* 134). In contrast to the large greyhound alongside the adult noble's body in *Cardinal Infante Don Fernando as Hunter*, the presence of the medium-sized *podenco* suggests the smaller size of its subject.[20]

The young prince died only a decade after Velázquez executed *Baltasar Carlos as Hunter* and a decade before *Las meninas.* Velázquez most likely remembered the untimely death of the young Habsburg heir Baltasar when he painted dogs in future child heir compositions. *Felipe Próspero* (1659) is a most illuminating Velázquez painting in understanding the dog's connection to the death of royal progeny (Figure 4.5). *Felipe Próspero*, the last significant portrait that Velázquez painted before his own death, shows that the artist depicted a dog in his examination of Philip IV's bleak future. Philip's son is two years old in the painting and will only live to be four. In other portraits of royal children with small dogs such as Johan Baptista van Uther's *Sigismund III* (1568) and François Clouet's *Hercule-François* (ca. 1556-58), the children are posing, dressed as potential monarchs. But in *Felipe Próspero* Velázquez depicts the young boy as dressed in a way that attempts to hold out against the impending arrival of the end. The child, known to be sickly, wears amulets as a defense against possible ills. The amulet in the shape of a fist and pointed finger (known as the *higa*) on his left shoulder served in defense against the evil eye.

---

18 Of German origin, historians still have not confirmed Maribárbola's original name and biographical profile. For a study of Maribárbola as important figure in Philip's court and not simply a dwarf figure that contrasted the figure of Margarita, see Messerli.

19 The *encarbo*, *encaramo* or *engarbo*, the early modern Spanish word in hunting manuals for a pointer, would later be identified as an Old Spanish pointer in the nineteenth century and cited as a distinctly autonomous breed. Breeders today call this uniquely Spanish breed a *perro pachón navarro* or *perro de punta ibérico*.

20 Even more popular than the dwarf-dog genre, an early modern portraiture tradition existed in Europe in which the size of the dog mimics the size of the sitter. Titian painted a small child with a small dog in *Clarice Strozzi* and Sofonisba Anguissola painted Isabel Clara Eugenia and Catalina Micaela, Philip II's daughters, with their little spaniel on a table (ca. 1561).

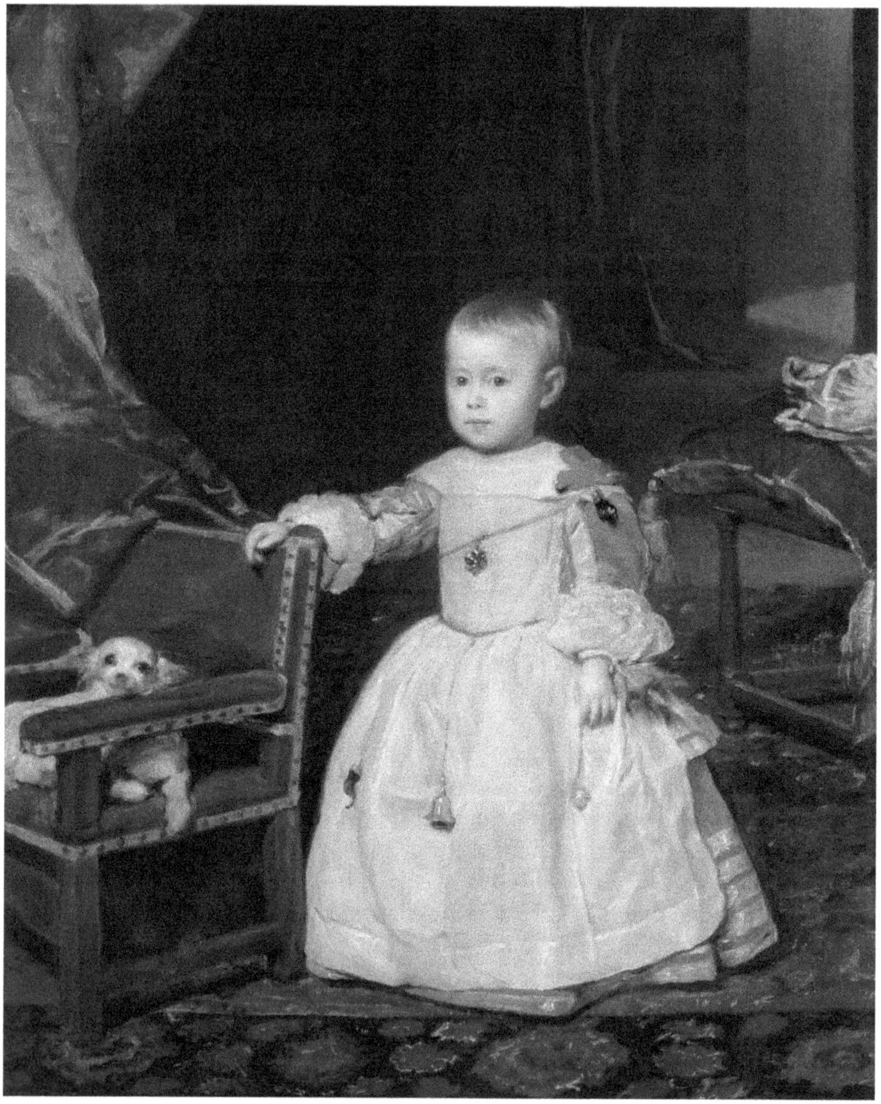

Fig. 4.5 Diego Velázquez, *Felipe Próspero*, 1659, Vienna, Kunsthistorisches Museum.

A spaniel accompanies the sickly effeminate Felipe who wears amulets to somehow trick death from stopping at this house. But the amulets ultimately serve no purpose. The child will never reach adulthood. The spaniel connects with the white figure of the child, a ghostly signal that Velázquez senses that death will enter Spanish royal quarters.[21] The enveloping darkness in the background of the painting finds materiality in the dog's jet black eyes. The dog has literally and symbolically taken over the royal quarters carelessly lounging on a chair that few bluebloods would sit on without pause given the rigid etiquette of the Habsburg court.[22]

The dog's other-worldly eyes have seen the other side of the abyss, a place that Felipe's amulets will never have the power to penetrate. If a child is a little angel and a sign of future life, then, since a dog has entered their house, this child angel will soon disappear to the land beyond. The portraits of Felipe Próspero and Margarita in *Las meninas* represent similar figures. Both have fair hair and wear elegant dresses. The left hand of each child rests on the side and each right hand is outstretched and open. Felipe, like Margarita, is a potential monarch, but also an image of a failed royal male heir.

Aside from their role as guard dogs, hunting animals, and pets, dogs in King Philip's court also suggested the general deteriorating conditions of every day life. They lived with the royal family and one text that describes court life in 1656 comments that a dog that, when given the opportunity to partake of some of the food the royals were eating, would not touch it because it stank and was clotted with flies (Paz y Melia, III 24; qtd. in Waite). Observing the dog in *Las meninas* in conversation with *Felipe Próspero* helps elucidate how Velázquez' associated the dog with the end of the future royal line. The dog is not a symbol of faithfulness, the prospering noble family portrait, and future family that started in the early modern period with such works as Gilles Van Tilborch's *Portrait of a Family*, Giuseppe Arcimboldo's *Maximillian II and his Family*, and Anthony van Dyck's *Children of Charles I*. In *Las meninas* the dog specifically connects to a growing preoccupation of the King. The dog does not sit by him in a signal of his masculinity and by extension new Habsburg progeny. Since Margarita is the central image in *Las meninas,* one also must imagine her relationship to the

[21] The dog also senses death in *Joseph's Bloody Coat* (1630), Velázquez's only other painting that, aside from *Felipe Próspero*, contains a small dog. In *Joseph's Bloody Coat*, the dog is depicted similarly to those paintings that Velázquez would have seen in Italy in which the dog was typically included in Christian sacred events and had a special intuitive and prophetic sense (Cohen). Velázquez was especially interested in the dog's ability to sense a lie about death since in *Joseph's Bloody Coat* the spaniel knows Joseph's brothers have come to deceive their father about Joseph's death.

[22] To explain the dog's attitude in *Felipe Próspero*, my daughter provides the following reference from *Good Dog* when the King Charles spaniel states: "WHAT ARE YOU DOING IN MY CHAIR!? / There, now that I've resumed my throne / perhaps you could fetch my royal rubber squeaky burger. / I'll forgive you this once, / but next time it's the doghouse for you" (Gottfried).

dog. Margarita stands under empty hooks, a sign of absent chandeliers (wax was expensive) and austerity. Moreover, even though many argue that she is being prepared for coronation and there was no Salic law in Spain (she could be queen), Margarita represents potential masculine shortcoming for Philip IV since she is a "she" and the end of his hopes for fulfilling a male heir. Geoffrey Waite writes with respect to Margarita: "Female children are quite a bitter disappointment for this court" (273).

The central image of Margarita highlights the way in which a preoccupation with the future of pure royal bloodlines pervades the fabric of the painting. The painting contains a central object, a little red pitcher or *búcaro*. The small red pitcher contained perfumed water and, after drinking, was eaten by aristocratic ladies in order to whiten their complexion. The eyes from the image of Velázquez, then, do not only look up to see the dog and the monarchs about to leave, but they also take in Margarita's soon-to-happen ceramic-eating act that will whiten her complexion even more.

The archeologist Natacha Seseña has undertaken the work that helps make the association between *Las meninas* and royal worry over *casta*. Seseña has researched the particular shape and color of the *búcaro* and has uncovered a plethora of sixteenth- and seventeenth-century texts in which references are made between eating the pitcher and making one's skin color white. Covarrubias also writes that consuming the *búcaro* acted as a whitening agent for female nobility: "ladies eat these clay containers to make themselves more pallid" ("destos barros dizen que comen las damas por amortiguar la color" [my trans; 239]). Aside from the red *búcaro* whose clay was most commonly brought from Tonalá (Guadalajara in Mexico), the painting also displays a silver tray and red curtains. The silver was most likely mined from a silver mine in the New World like Zacatecas and the red curtains in the mirror reflection over the heads of the monarchs were dyed with American cochineal (Ellsworth Hamann). The dog had been a key tool and icon in the conquest of America and the American material objects bring the invisible reflection of Spain's imperial conquest to the painting. Linked to a failing American enterprise, the objects are part of an offering to Margarita, Philip's failed attempt at a male heir, and they, like the dog, signal to the king the prospect of an uncertain family future.

The Spanish mastiff in *Las meninas* indirectly evoked the future family demise of the Habsburg line. As the seventeenth century advanced, the connotations of Spanish mastiff were less that of a majestic animal and more that of an impure and ugly mix. The dog's placement next to the king who poses for a portrait is *not* represented next to the king in visual representation of the painting. This canine separation from the king evokes a ruined royal friendship with the dog. Indeed the king-dog relationship in the painting suggests the Habsburgs may no longer be *para casta*, that is, the king and his progeny may not be pedigreed. In his sixteenth-century husbandry manual, Herrera describes the pure bred stock of the Spanish mastiff using the word *casta* insisting that it should be bread *para casta* and of *buena casta* (348). In contrast, in the seventeenth century when Velázquez was painting *Las meninas,* some writers began to question the dog's *casta* (its purity) casting doubt on its sixteenth-century noble reputation. The mixed origins of the Spanish mastiff had once given it a fearsome and awesome pedigree, but, as the seventeenth century advanced, the Spanish mastiff became increasingly associated with mongrel dogs and became a synonym of the outsider. Even though the etymology of Spanish mastiff is *manso* (tame), seventeenth-century lexicographers connected the word with *mixtus* (Covarrubias 793). In one play from the period, *Juan Latino*, a white character mocks the black protagonist by calling him a Spanish mastiff. By the early eighteenth century, in the 1734 installment of the *Diccionario de Autoridades* (the first official dictionary of the Royal Academy of Language), "Spanish mastiff" would come to mean a monstrous, ugly human: "One can compare this dog to the perversely rough and crude man; because of this, the expression 'That guy is a Spanish mastiff.'" ("Por comparación se llama el hombre feamente robusto y tosco: y así se dice, 'Fulano es un mastín'" [my trans.; 509].

In *Las meninas,* Velázquez recognized various canine iconographic traditions associated with the representation of the humanist, dwarfs and the nobleman and his family. But Velázquez primarily associates the dog on the threshold of an approaching death, particularly as related to the King's military, economic, procreative power, and family reputation. Beyond the classical positioning of the dog as a threshold figure between the world of the living and the dead, the European medieval tradition associated the dog with death in a variety of ways: tombs represented dogs resting at the feet of men and women; medieval bestiaries depicted dogs as living in the sepulchers feeding on dead remains (White 31); and artists painted dogs as emblematic of death. For instance, paintings connected the dog to the theme of *vanitas* and the brevity of life. In the fifteenth century Hans Memling painted a triptych, *Earthly Vanity and Divine Salvation*, in which the left panel shows an image of a skeleton and the right one shows the devil. The central panel portrays a nude woman looking at herself in a mirror with dogs at her feet. The fluffy griffon-type dog at the left and the amorous greyhounds at the right suggest physical love and ornamentation. The small dog and the greyhounds are placed on a beautiful field of flowers, and they all, like the woman's beauty, are fleeting, false forms of beauty. In this way the dogs and the other pretty images are an artificial lure for the left and right panels that show a skeleton and hell.

In contrast to the representations of humanists and nobles with dogs, a marginal early modern painterly tradition continued the medieval association between the dog and death. Two sixteenth-century paintings had an important influence on the creation of the dog in *Las meninas.* Fernando Marías is one of the most important Spanish scholars of *Las meninas.* Marías notes that Tintoretto's *Christ Washing the Feet of His Disciples* (c.1556; this painting hangs today in El Prado) (Figure 4.6) is an important source painting for Velázquez (248). Velázquez would have seen *Christ Washing the Feet of His Disciples* in the Sacristy of the Monastery of the Escorial in 1656 (it was part of the Spanish royal collection since 1654). Velázquez would have also carefully studied the original in Saint Mark's Basilica

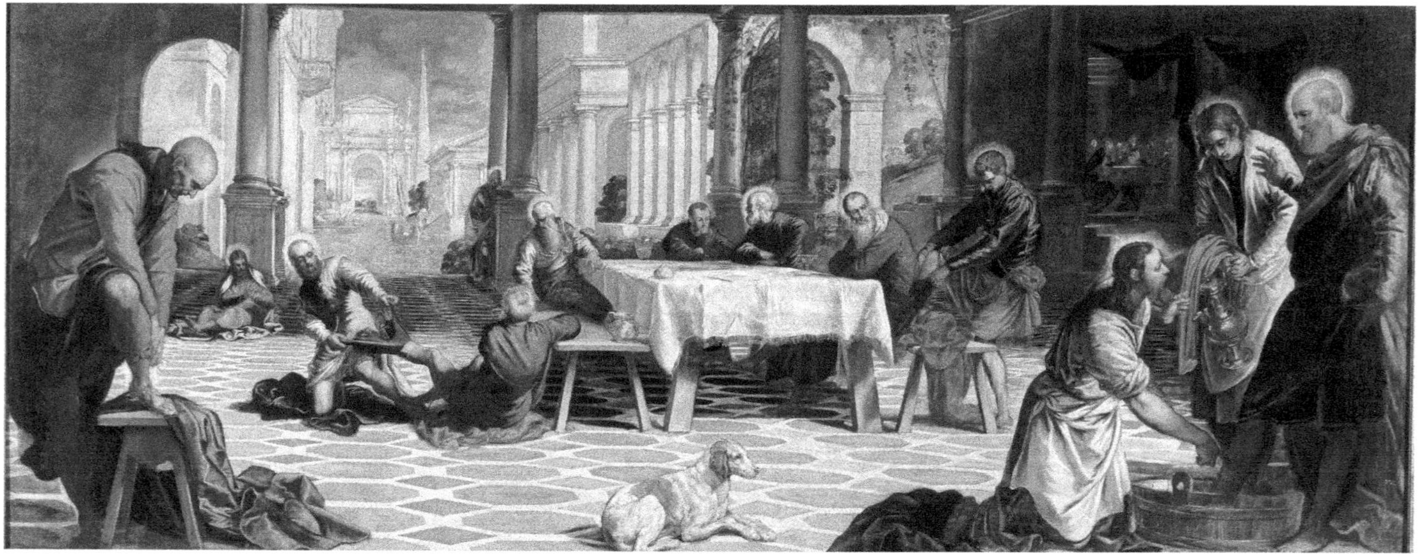

Fig. 4.6 Tintoretto, *Christ Washing the Feet of His Disciples*. c. 1545–55, Madrid, Prado.

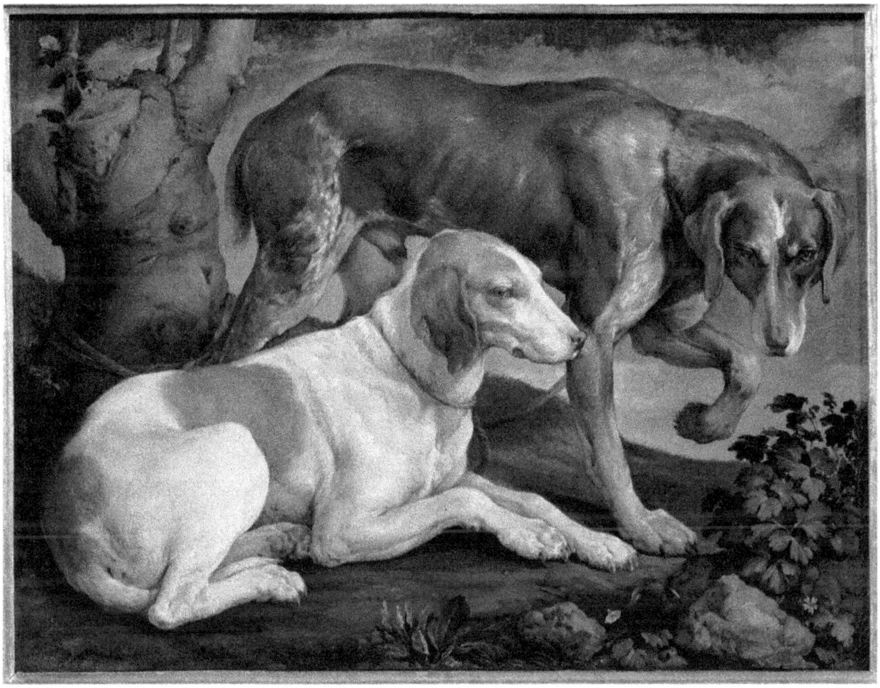

Fig. 4.7 Bassano, *Two Hunting Dogs*, c. 1548–50, Paris, Louvre. Photo Credit: Erich Lessing/Art Resource, New York.

in Venice (Marías 248). But Velázquez does not look to Tintorreto's dog in *Christ Washing the Feet of His Disciples* for the dog in *Las meninas.* His depiction of a dog was more influenced by Bassano. Tintoretto had copied the dog in *Christ Washing the Feet of His Disciples* directly from the identical dog in Bassano's *Two Hunting Dogs*, a painting that Velázquez would also have seen in Venice (Figure 4.7). Bassano's *Two Hunting Dogs* is dedicated to dogs and to dogs alone (Bowron 11-12). The rope in it signals an absent human, also signaling that a dog's existence is characterized by *waiting*. Dogs wait just as death waits. No animal has the capacity for waiting for the human as much as the dog. The incredible capacity that dogs have to wait has perhaps inspired many cultures, including Velázquez, to suggest a dog that never tires from waiting for the living.

In contrast to Marías, the North American art historian Jonathan Brown (*Las meninas*) argues that Rembrandt's *Night Watch* (1642) (Figure 4.8) is the most significant painting that influenced the creation of *Las meninas* because it captures a snapshot of an ongoing action. Like Marías, Brown's analysis of *Las meninas* ignores the presence of a dog in the source painting. As in Bassano's *Two Hunting Dogs* (the source for the dog in Tintoretto's *Christ Washing the Feet of His Disciples* and, in turn, an important source for *Las meninas*), Rembrandt's *Night Watch* also contains a dog ignored by *Las meninas* art critics. The dog is a haunting

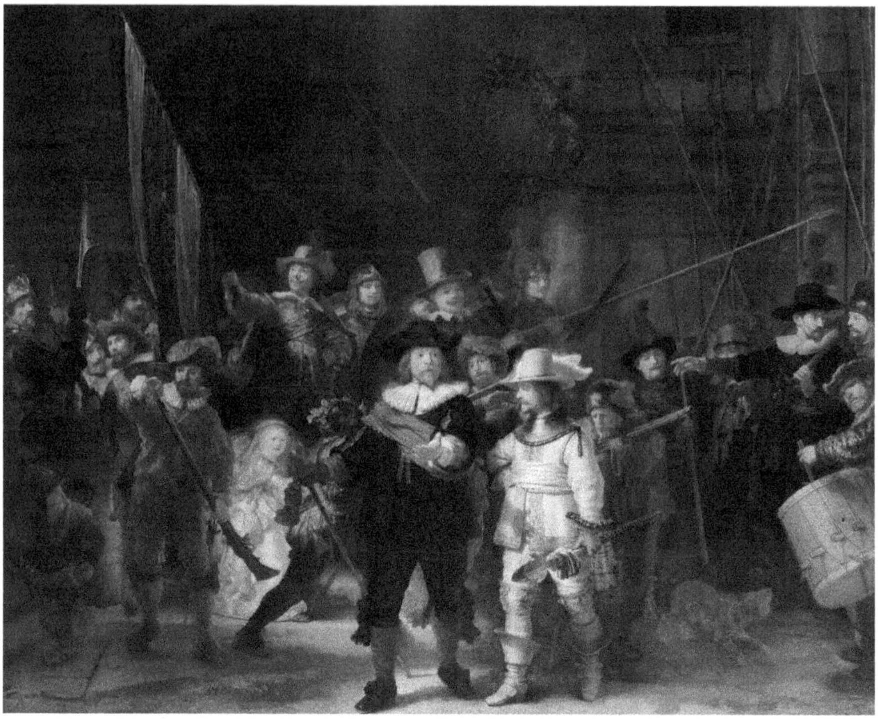

Fig. 4.8 Rembrandt, *Night Watch (The Militia Company of Captain Frans Banning Cocq and of Lieutenant Willem van Ruytenburgh)*, 1642, Amsterdam, Rijksmuseum.

image: a darkened, crouched figure in the shadows almost wavering above the painting's surface at the right front.

Western societies since the Greeks have had an ambiguous relationship with the dog when it comes to death and the ritual of burial. On one hand, in the Greek, Roman, and renaissance traditions, the dog was given a proper burial.[23] In this sense it was made human. In early modern Spain, the author of the poem, "El Alano," mourns León's death and writes that it deserves a mausoleum: "In Molossia they did not act without reason / when they nobly buried their dogs" ("No sin razón

---

[23] Take for instance the following ancient Greek elegy by Simonides: "Although beneath this grave-mound thy white bones now are lying, / Surely, my huntress Lycas, the wild things dread thee still" (Bowra 199). Catherine Johns includes a beautiful image of dog Latin epitaph and tombstone for the dog Margarita from 1-2 CE (198). The epitaph and mausoleum to the dog was revived in the early modern period such as the humanist Lipsius' mausoleum to his dog (Papy).

Molosia por costumbre / Noblemente sus perros sepultaba" [my trans.; 9]).[24] On the other hand, dogs were represented as the vermin that desecrated the unburied body and, indeed, were the "animal" that made human burial necessary. In the classical tradition the dog often commits the greatest abomination by devouring the dead before proper burial. One Renaissance philosopher argued that only humans bury their dead, making the ritual of burial a sign of the omnicompetence of the human.[25] In the Spanish picaresque novel *The Life of Justina, the Country Jilt* (*La pícara Justina*), the main character sardonically shows her father has no virtue and should not be looked at as a dignified human, after recounting that a dog has eaten his remains. The dog leaves the body "dogged" or "all messed up" ("emperrado") after "slashing him and slicing up his body and face" ("diole de tajo y destaxole el cuerpo y cara" [my trans.; 118]).

The tradition of the noble and humanist dog portrait reflected the spirit of the practice in which the body of each was projected into and out from his dog. The nobleman and the humanist thereby often gave their dog a dignified "human" burial. *Night Watch*, in turn, captures an alternative mindset with regards to dogs and death. It is a military portrait that captures the action in motion as a company of civilian militia moves out to the battlefield. The dog does not stand by a noble or humanist figure, but crouches in the shadows and very well may be the animal that does not let some dead soldiers have a proper burial. Will this dog delight in rolling in carrion and gnawing at abandoned corpses on the battlefield? The *Night Watch* dog is not represented as beloved and faithful to a noble master, but serves as a palimsest for the slumbering *mastín* in *Las meninas*, a reminder, like the white lap dog in *Felipe Próspero*, of death.

Aside from the medieval tradition and Bassano and Rembrandt's dogs that suggested an association with death, the early modern Spanish poet Antonio de Solís and the Spanish emblem of the cynocephalus provide two important models of dogs that serve to elucidate the connection between death and the slumbering mastiff in *Las meninas*. Antonio de Solís wrote a poetic anecdote to show how dogs had a special ability to recognize the power of the king. Antonio de Solís indicates that *smart* dogs knew to lie next to the king. His poem describes a celebration on the street that the king attended and a dog came up and sat at Philip IV's feet even though all his attendants tried to shoo it away: "He stops when he arrives to your Highness, / and he lies down snuggled up / against your feet, not leaving

[24] Molosia—the land where the Molossier is reputed to have come from and the place that would be used by nineteenth-century dog breeders to coin the Molossian dog variety—also interred its dogs. Dog manuals today still classify the *alano* as Molossier.

[25] The first lines of the Homer's *Iliad* describe bodies that are condemned to Hades because they were left unburied on the battlefield and eaten by birds of prey and dogs. The Renaissance philosopher, Giambattista Vico, asserted the omnicompetence of the human by arguing that only humans bury their dead. Vico's connection between the human and burial depends on a deep-rooted human-animal separation from the Biblical and classical tradition with respect to the dead body—the interdiction that the human body should not be consumed by the animal (Harrison xi).

them" ("En llegando a Vos, se para, / Y acorrucado se postra / A esas plantas, sin dejarlas" [my trans.; 120]). The dog *should* lie down next to the King's feet and in *Las meninas* it may have been lying next to the King's feet when Velázquez composed the picture, but *Las meninas* separates the painted portrait of the dog from the king.

In another poem, Solís shows one important way in which the dog connects with death. This poem is ostensibly about a triangle of entangled love: a poet, his lover, and a dog. Ultimately, the dog wins eternal affection of the poet's lover, not letting the poet gain her love. It arrives to his mistress and, like Odysseus' dog Argos, dies when he reaches her. The last lines of the Solís sonnet state that the dog that died by his lover's side made the best choice in securing eternity: "The dog made a rational choice in the turn of events, / Since the one that knows how to die at a fortunate time / Also knows how to eternalize happiness" ("Racional elección tuvo en la suerte; / Que quien sabe morir, cuando es dichoso, / Sabe también eternizar la dicha" [my trans.; 118]). The dog in the Solís sonnet is like the lover who hopes he can at least be the shadow of his lover's dog in Jacques Brel's famous song—"Let me become … The shadow of your dog" ("Laisse-moi devenir … L'ombre de ton chien"). Although the dog does not die in *Las meninas* as in the Solís poem, the poem reveals a key purpose of the dog in *Las meninas.* The dog is not the poet's companion, just as he is not the King's companion. The poem suggests that the dog is aware of death, but that his master will never fully perceive it. Might the dog in *Las Meninas* be aware of death too, but his master, King Philip, never fully perceive it? Indeed, the call to the impossibility of knowing death ensures that King Philip, like the poetic persona of this sonnet, always arrive too late.

When Velázquez set out to paint the dog in *Las meninas,* he studied the detailed anatomy of the Spanish mastiff and the animal in the painting is a remarkably realistic reproduction of a dog. Nonetheless, Velázquez did not paint the dog within a traditional humanist framework that understood the omnicompetence of humans as the highest members in the chain of being. The dog in the painting signals a chiasmic interplay between the animal and the human and between the living and the dead. Even though the anatomy of the dog is precise, a description of the cynocephalus from early modern books of emblems provides a better model and hermeneutic to understand the Velázquez dog than the many realistic images of dogs from the period.

Books of emblems were quite popular in the Spain and were important influential sources for Velázquez (Moffitt). In Juan de Borja's *Empresas morales* (1581), one of the emblems is entitled "We Die a Little Each Day" ("Quotidie Morimur," [390–91; Nelson 50–51]). The emblem is an image of a slumbering human figure with a dog's head. (See the cover of Nelson's book for an illustration.) As mentioned in Chapter 1, the cynacephalus had long been conceived as belonging to a monstrous race of creatures on the margins of the inhabited world. It also had other legacies, including that of a living-dead being. Borja provides a gloss for the emblem, explaining that a cynocephalus is somewhere between a human and an animal ("gente o animal" [390]).[26] The human–dog creature lives *and* dies. Borja writes that the cynocephalus "dies little by little, with its foot dying one day, the next day its hand, and in this way it ends its life" ("muere poco a poco, muriéndosele un día un pie, y otro día una mano, y así se le va acabando la vida" [my trans.; 390]).[27]

The dog, however, does not follow the emblematic tradition that pre-supposed the medieval bestiary idea that the world was a great cryptogram and the image of the animal awaited an interpretation that unlocked its mystery. Nor is the dog in the *Las meninas* purely an image of a dog that is to be studied for its realistic proportions to be classified in a Linnean taxonomic system. The "in-betweenness" of the *Las meninas* dog pushes one to consider how death approaches the world of the living and a live body approaches the world of the dead. The dog straddles the threshold between both. Its front paws, almost touching the bottom of the painting, reach toward the outside of the front of the painting, constituting the spectators' first bodily entrance in the painting. In a similar way that seventeenth-century palatial plays displayed animals on the proscenium and curtains, so the dog is a liminal body for Velázquez's pictorial stage. In one palatial stage from the period, an image of a lion is placed on the threshold of the proscenium (see Plate 1 of Greer, *The Play* for an illustration).

The pictorial stage of *Las meninas* separates the figures within the painting from both the king and queen outside the frame. *Las meninas* at once attests to the royal couple's *live* presence alongside one of his dogs, but also their absence. The king's *real* portrait is nowhere to be found next to the dog; he is a phantom, a living presence that only exists in *Las meninas* alongside the queen in a fuzzy mirror image at the back of the room. The paradox and liminality of the dog may be expressed in the following way: the dog's formal placement in the foreground separates the living king from the represented images in the painting.[28] The dog thus separates the living figures represented inside the painting from the king's deathly nothingness outside the painted surface.[29]

---

26 As an aside, for further reflection, I would mention two images that show an alternative version of the hybrid creature with a dog's head. These two images illustrate a dog with human ears. One can be found in an eighteenth-century illustration of "The Dialogue of the Dogs" in which the dogs have human ears (Manning 140). The other can be found in an American codex in which the god Xolotl takes a human form with dog's ears (Keber).

27 For the medieval tradition of the cynocephalus, see Schmitt 148–9.

28 Like the Ojibwa legend it is as though the dog in *Las meninas* crosses over to the other side, the side on which we stand and the king and queen stand. The king, alive, is separated from the dead painted image *and* the painted dog is forever alive, separated from the dead king in the frame of a mirror and who observes the painting from the outside of the canvas.

29 A Picasso and Goya dog suggestively align with the canine's association with death and liminality in *Las meninas.* A century after *Las meninas,* Goya painted a lone dog on a barren landscape. Goya's dog may be about to appear from behind a hill or it may be about to disappear. Pablo Picasso, in his final full-scale study of *Las meninas* (1957), was concerned with the dog. Picasso eliminates the slumbering Spanish mastiff and replaces it with his dachshund "Lump." Lump slept with Picasso and shared dinner with Picasso,

Velázquez looks to a dog that is a neither human nor animal that tells the king of an approaching death. The dog in *Las meninas* is not a friend, that is, an appendage to the humanist that discovers truths that elevate humans toward the divine, nor a royal portrait companion that shows off the king's masculinity, faithfulness, and contained power. The dog reminds the king of certain truths that increasingly no longer hold, such as the central role of humanism in thinking toward the truth. It also leads him to question his masculinity, the future of his progeny, and his empire's military prowess and economic condition.

In examining the way in which Velázquez represents the dog, one can point to some curious parallels with Cervantes. Cervantes and Velázquez deviate from the sixteenth-century humanist-dog tradition in which the humanist sits by a slumbering dog (and they also deviate from the later permutations of the humanist-dog tradition in which the humanist is an animal![30]). They both represent the

---

eating off his dinner plate. Picasso understood the dog as a unique other. He wrote about Lump: "Lump is not a dog, not a man—he is really something else" (Duncan). Picasso calls the dog "something else" and makes it a white blotch of a dog. The white blotch represents a living dog with a name, Lump, but it is also a ghost of a dog.

[30] One can find permutations as of the humanist-dog tradition in which the humanist is an animal in Hogarth, Restif, and Bass. In 1749, William Hogarth, the English painter and early animal rights' activist engraved an image in which his framed round self-portrait is placed on top of tomes by Swift, Shakespeare and Milton (Johns 140). In Hogarth's meta-painting, the image of the artist is not "real," but only a painting, an artifice, because it is contained within a frame. In contrast, Hogarth places his pet pug Trump in the foreground as a real being; the dog sits just in front of the painting of Hogarth's face that sits on top of a pile of books. The "humanist" in this case is a dog since the artificial props, the artist and the books, provide flesh and truth to the animal, Trump. In contrast to Hogarth's framed image of his self-portrait, Velázquez does not let his viewer see what he is painting on the canvass. Moreover, the self-portrait of the artist and the dog in *Las meninas* are represented on the same plane in the reality of the narrative the painting, not in contradistinction to each other. Nicolas-Edmé Restif de la Bretonne also takes the humanist-dog tradition and reverses its terms. In Restif's novel, *The Southern Discovery by a Flying Man, or The French Dedalus* (*La découverte australe par un homme-volant, ou Le dédale français*, Leipzig, 1781), the humanist is literally an animal, the cultured baboon, César de Malaca. In one illustration of the book, César is writing with a dog at his feet. The humanist baboon-dog illustration includes César's words: "Thank God that I am an ape and not subject to human laws" (qtd. in Schiebinger 109–10). In 1999, Rick Bass writes a memoir and elegy for his favorite hunting dog Colter, a German shorthaired pointer, who lived with him in Montana in Yaak Valley. The reality of the sleeping dog's body contrasts the shadowy, unreal world of Bass's writing: "Looking in on that sleeping dog at night, after a good hunt, before going to my own bed to turn in for the night, en route to the next morning's slow seated work, it would seem always and again that one of those two elements, the dog, was always real, and the other, the writing, always shadow" (17). Bass places the dog in the world of the truth or reality in a contrast to the tradition of the "human" humanist whose scholarship and findings were truth and reality. But in Cervantes and Velázquez, neither the artist nor the dog is more "real" or more "shadowy" to use Bass's terms. Hogarth, Bass, and Restif reverse the human-animal terms of exceptionalism: the animal is exceptional at the

animal as occupying a place outside of human *and* animal exceptionalism. In this way, each avoids the perpetuation of an animal–human hierarchical entrenchment.

Cervantes and Velázquez do not design the dog as an ennobling symbol. The dethroned "lion" dog in Velázquez's *Las meninas* compares with the passive, less-than-fierce lion in *Don Quijote*.[31] Picking up on Pliny's assertion that lion can move its tail like a dog and the tradition of dog-like friendly lions in books of chivalry, the lion represented in Part II of *Don Quijote* transforms in a sort of dog that does not bite or "even bark" (Hart 21; Chiong Rivero 1023). The dog in *Las meninas* and Velázquez's other portraits—more than a noble emblem—better suggests a defeated spirit not unlike that which can be found in the first sentence of *Don Quijote* in which the aging knight has an old shield, a skinny nag, and, significantly, a hunting hound. Although a different historical period (*Don Quijote* is 1605, three decades earlier), the dog that Quijote will abandon from the first sentence of Cervantes' novel suggests a similar dejected spirit that can be found Velázquez's noble-dog images in which broken dreams cloud the hunt and augur an ended glorious age.

Velázquez and Cervantes both study the slumbering dog and explore the question of whether sleeping dogs should lie. Pertusato does not let the dog in *Las meninas* lie, just as the madman does not let it in the Prologue to Part II of *Don Quijote.* The image of the dog lying at the feet of the people in *Las meninas* parallels the image of Berganza and Cipión at the foot of Campuzano's bed in the hospital. The *Las meninas* dog—an image found in the King's private quarters according to art historians—probably inspired King Philip IV to contemplate his life. Moreover, in the fictional Cervantine world, the two lying dogs in "The Dialogue of the Dogs" accompanied Campuzano in his recovery to physical and mental health, influencing his movement forward in life.

Human dependence on and interdependence from dogs are an important part of representation in the artistic projects of Cervantes and Velázquez. Dogs appear in the work of both as each considers his own mortality. The most significant canine connection between Cervantes and Velázquez is the fact that both men choose to design dogs as death approaches. In *Las meninas,* a late work, and *Felipe Próspero,* the last important portrait he paints before his own death, Diego Velázquez studied and painted a dog. Toward the end of Miguel de Cervantes' life, he also turned to the dog. In the Prologue to Part II of *Don Quijote,* one of the last works that he wrote before he died, Cervantes wrote two fascinating canine anecdotes.

---

expense of the human and humanist thought. In contrast to these three examples of animal exceptionalism, the dog in Cervantes and Velázquez is not humanized while the humans are less "real" or more brutish.

[31] For another European example, specifically a commentary on the case of the motif of the African lion turned into a little "lion" lap dog in fifteenth-century France (that manifests desire with psychological, social and political consequences), see Schiersari, "Bitches."

Afterword
Amores perros

One may argue that the point of the film *Amores perros* or *Love's a Bitch* (the official translation of the title) (2000) is to depict the sad, animal state to which some people in Mexico City have descended. The argument, following traditional interpretations of the animal as figure for the human, maintains that the characters in *Amores perros* become "animal" in the sense that they care more about their dogs than human beings. The three narratives in the film focus on the three protagonists, Valeria, Martín and Octavio. Based on this interpretation, the title, *Messed-up Loves* (the literal translation of the title), refers to each of these three characters' unhealthy affections toward dogs. The model Valeria fondles her white lap dog as if it were her baby. Valeria loses her leg and her dog in the film. The prosthetic leg is a sign of how Valeria's body, her legs, made her an object of value in this economy. But the prosthetic leg also signals the missing little white dog since dogs in the world of *Amores perros* are appendages, creatures that are strangely not quite human, but at the same time an indispensable part of the body of these three protagonists.

The scene of a car accident when Valeria loses her leg climactically brings her together with the other two protagonists, uniting the three vignettes that are otherwise separate filmic narratives. The second character, Octavio, crashes his car into Valeria's. Octavio, escaping from a scene of violence at a dog fight, desperately wants to save his beloved companion Cofi, a bull dog mix, who lies shot and bleeding in his backseat. The third character Martín Ezquerra (El Chivo) mercilessly kills other men for money, but shares his run-down abandoned flat with a pack of stray dogs that he cares for with tenderness. When Martín Ezquerra arrives to the scene of Valeria and Octavio's accident, he ignores the suffering people, but carries away Cofi to his flat and he nurses the dog back to health.

Even though it is an important theme, the ultimate premise of the film is *not* to depict an "animal" or "dog-like" state of people who care for animals more than people. In the final scene of *Amores perros*, the director of the film Alejandro González Iñárritu presents a radically new conception of the "messed-up love" (*amor perro*) through a redeeming reconfiguration of brotherly and animal love. This final scene highlights the interdependency of the character Chivo and the animal Cofi by displaying two beings that are not friends, but nonetheless, companions on a journey. El Chivo walks into a barren wasteland in a newly-forged animal relationship in which the dog is not a human appendage, but, as Derrida would call it, an "other other."

The film evokes a new story of Abraham and Isaac because of the way in which it conceives of the animal in the logic of sacrifice.[1] With his pack of dogs slaughtered by Cofi, his new master Chivo stuck a gun to the dog Cofi's head and, like Abraham's near sacrifice of Isaac, almost killed him. Cofi is spared like Isaac, but he is not an "other" like Isaac because, after the near sacrifice, Chivo does not choose a replacement animal, a scapegoat (*chivo expiatorio*), to sacrifice in lieu of Cofi; the potential sacrifice victim already *was* an animal and it is spared.

Moreover, this portrayal of Cofi, the animal who was to be sacrificed, also reconfigures the relationship logic of brotherly love found in the story of Cain and Abel. Cain, angered because God favors Abel's animal sacrifice over his of wheat, seeks out and murders his brother. The director González Iñárritu uses the moment of Chivo sticking his gun to the dog's head to connect to the theme of brotherly hatred in the film. Octavio's brother had also put his gun to his brother's head.

If Chivo were to have killed Cofi, he would have sacrificed the animal. But the final scene of the film does not focus on biological brothers, but on a newly-forged "brotherly" relationship between Chivo and Cofi. One critic writes that the final scene is "the only time the horizon is seen in the film" and that the ominously flat landscape of the scene "lies beyond human habitation, which is revealed by default as the unique possibility for love" (Smith 57). This love is a unique sort of brotherly and animal love. In the Cain and Abel story, God desires the sacrificed animal, causing one brother to murder the other, and then the banishment of the murderous brother. The film makes the viewer remember that this dog, unlike the animal in the Abel-Cain story, is *not* sacrificed. The final image of the film shows two beings whose love—whose *amor perro*—is outside of the legacy of Cain and Abel, that is, the sacrifice of the animal that triggers the legacy brotherly conflict and killing.

The final scene of the film looks beyond the "messed-up" loves in the film to a meaning of "love of dogs" quite different from Valeria and Octavio's mutual canine dependency and the canine dependency that Chivo had when he lived in a house filled with strays. The final suggested meaning of *amores perros* exists in the land of the love for the other other, a new sort of love not based on the cliché of "man's best friend" and a far cry from a Lassie home. When El Chivo leaves Mexico City with Cofi into a wasteland, he walks into an Agamben naked life. El Chivo does not abandon the human to become animal nor abandon the animal to become human. The two companions walk into a living death, forcibly banished like Cain. This exile is different than Cain's, though, because it marks the end of a plagued brotherhood that stems from animal sacrifice, that is, an ethics that depends on a notion of sacrifice in which the semantics of animality create the human.

---

1 Martín's nickname Chivo (literally "goat," but also the one who reports a person's misdeeds to another) reminds the viewer of sacrifice since in Spanish the *chivo expiatorio* is the sacrificial victim or scapegoat.

One may take this Animal Studies reading of the final scene of the film as a way to read the animal in Cervantes and Velázquez in the sense that the critic *can* perform a traditional reading of the animal in their work by citing how the animal is simply a mask or figure for the human. But the portrayal of the dog in the film, as in the work of Cervantes and Velázquez, is most productively interpreted through an Animal Studies lens. Moreover, in assessing whether there is historical continuity or a significant break between the role of the dog in *Amores perros* versus the case of the early modern Spanish world, I would conclude the following: On one hand, a continuity exists between of Velázquez and Cervantes and films such as this one since they share some points of contact in their treatment of the animal. On the other, the animal was rarely the cynosure of the Spanish early modern period. Taking the contrast between *Amores perros* and *The History of New Mexico,* for instance, we could say that in the film the dog breathes interdependent of its master. Cofi is not left nameless to die unburied like Pérez de Villagrá's dog, but walks, living, into the future.

Appendix
The Animal in Identity Categories

In the sixteenth-century Spanish world, *raza* and *casta* were used interchangeably for breeds, species and lineages for animals, plants, and humans. Eighteenth-century European scientific thinking from outside Spain largely influenced the usage and meaning of the word *raza* as "race" and "breed" in contemporary Spanish (and the disappearance of the word *casta*). Despite the post-Enlightenment semantic evolution of the word *raza*, scholars have not generally acknowledged that pre-Enlightenment Spain, largely due to its American experience with animals, contributed *to* the formation of European thinking on race. Spain, then, before it was influenced *by* European thought on race, first had influenced Europe and contributed *to* the formation of European thinking on race.

The fear of out-of-control and impure animals directly linked the creation of race in the Spanish Atlantic mindset. When domesticated European animals reached American soil, many would reproduce out of supervised control. The fear of the breeding of the unregulated animal was tied to the formation of key racial identity terms, such as Creole or *criollo* and Maroon or *cimarrón.* Other human racial identity terms originated out of the notion of impurity of animal *casta*—a notion based on long-established husbandry practices related to the establishment of purity of breed. Those racial terms, such as Mulatto or *mulato* in the case of horses and *mestizo* and *cholo* in the case of dogs, originally meant impure and not fully complete examples of those specimens. These identity categories make manifest that the definition of not just race, but humanity, became contingent on the story of making one's own singular species "human" at the expense of silencing an outsider as animal (for example, the labeling of a people as "dogs" or "mongrels"). The sign of the animal, particularly the dominated and domestic animal, was also paradoxically made the sign of the insider. The genealogy of identity terms from the sixteenth-century Hispanic world illuminates how people that are othered as animal later incorporate and forget the sign of the animal, producing an unproductive anthropocentric loop in identity thinking wherein the sign of the animal is a disparaging racial term and also turns into a term of racial pride.

The etymological root of six identity categories reflects the zoological-human shift that occurred with the meanings of caste (*casta*) that occurred in the sixteenth century. The expression *para casta* was a term originally used in animal breeding and it basically means "to reach purity of breed." The early sixteenth-century manual, *Art of Agriculture* (*Obra de agricultura*, 1513), written by Gabriel Alonso de Herrera, uses the term for the selection of Spanish mastiffs in the creation of a pedigree specimen: "The breeder should not let any dog, stud or bitch, that he wants *para casta* unite before it is a year old because they will lose their value and strength" ("A ningún perro, macho, ni hembra, de los que quieren para casta no los dejen juntar, o tomarse antes que hayan año, porque desmedran mucho y pierden la fuerza" [my trans.; 348]). The logic of *para casta* from medieval husbandry practice connects to the notion of *castizo,* the adjectival form of *casta,* a term that suggests purity of breed. Generally, early modern dictionary writers—borrowing on its medieval importance—choose the horse as the exemplary animal for showing purity of breed.

Just as he overlooked the importance of the canine meaning of the Provençal word *espagnol*, so Américo Castro did not consider the zoological inflection of the word *casta*, a term that he extensively discussed especially as it related to the peoples of the three dominant religions in the Spanish Middle Ages. Drawing from its medieval prestige-granting power, some sixteenth-century Europeans turned to the collection of horses in which the *castizo* animal heightened its owner's status as *castizo*. The medieval practice of horse breeding had given prestige to the aristocrat. Equestrianism had given prestige to soldier (the origin of *caballero*). Moreover, horse collection in the early modern period bestowed a sense of symbolic purity to the collector. The humanist, Gonzalo Argote de Molina, (1548–98), who wrote poetry, was a historian, former soldier, bibliophile, a man of letters (he edited *El Conde Lucanor*), and a proto-curator. His museum, considered the finest in Europe, contained live horses.[1] Molina's practice of collecting purebred horses reflects the early modern notion that the horse was the primary zoological example in showing off purity of breed and, by extension, the social display of one's nobility and fine breeding.

Cervantes would mock the notion of *para casta* when its meaning shifted to human categories.[2] At the end of the Cave of Montesinos episode in Part II of *Don Quijote*, Sancho makes a reference to the husbandry practice of bettering one's animal stock by suggesting that Quijote has been in the cave so long that he has started his own family lineage. Sancho states that Quijote has been in the Cave of Montesinos *para casta*: "A very hearty welcome to your grace, Señor, we thought you were going to stay down there *para casta*" ("Sea vuestra merced muy bien vulto, señor mío; que ya pensábamos que se quedaba allá para casta" [603; 726]).[3]

The meaning of *casta* would be increasingly applied to human contexts throughout the early modern period. Just as it sought to control the breeding of its animals on its American territory, so Spain projected the zoological vocabulary of

---

1 It also contained weaponry, paintings and taxidermy animal specimens (including an American armadillo). For more information on Argote de Molina and his collections, see Palma Chaguaceda, *Millares Carlo*, and Checa.

2 Eighteenth-century dictionaries also cite the word *casta* as variously referring to a noble lineage of knights and horses (Alves 88). Dictionaries from the sixteenth and seventeenth century linked the word *castizo* to those families that have many children and in which both parents come from, as Mensheu (1599) puts it, a "good race" (Nieto Jiménez and Alvar Ezquerra 3: 2254). Mensheu's translates *castizo* as referring to healthy familial bloodlines. The same meaning for *castizo* is found in Covarrubias' Spanish dictionary: "We call those of good lineage and *casta* 'castizos'" ("Castizos llamamos a los que son de buen linage y casta" [my trans; 316]).

3 I have slightly modified Edith Grossman's translation. She translates *para casta* as "to start a family."

casta from animal breeding into its desires to control new colonial populations. In the 1550s-colonial context, when referring to human classifications, *casta* referred to racial difference since it not only referred to pure bred animals (*animales castizos*), but it also took on the newly-inflected human legal register, referring to different racial categories in the New World. As a precursor to the *casta* paintings that represented a variety of racial combinations in the make-up of families in the New World, *casta* books became prevalent in New Spain at the beginning of the 1600s (Martínez 146–7). In recording families of different racial compositions, *casta* books in New Spain implied the necessity of the control of the sexuality of women for hierarchical colonial social order, specifically that they remain chaste (*casta* simultaneously means not only "race" and "breed," but also "chaste"). The *casta* book in New Spain shifted the goal of the medieval animal husbandry to the human being. The notion of the production of a pure *casta* animal—the best-equipped animal created through the strict control of the breeder—was applied to the colonial human subject.

In the seventeenth and eighteenth century, often using terms that referred to animals, colonial artists increasingly produced *casta* paintings that marked and coined the names of a variety of others, using the categories of *español* and *castizo* as the differentiating marker of purity. Those *casta* paintings would be collected by artists across Europe, contributing to the notion a sense of race based on the Spanish notion of *casta.* By the eighteenth and nineteenth century in the Spanish-speaking world, the term *castizo* entered the inside-outside logic of national and racial identity formations, connoting racial and national pride in one's identity, culminating into a sign of Catholic masculine heritage in twentieth-century fascist discourse in Spain.

This fascist heritage developed out of nineteenth-century semantics in which the terms *castizo* and *raza* interchangeably indicated a pure, unsoiled Spanish identity. Like the semantic shift from *castizo* to a term of human identity and pride, a process of ennobling whose roots are zoological takes place in the trajectory of the etymologies of Mulatto, Creole, Maroon, Mustang, Mestizo and Cholo. Each of these contemporary English words finds its origin in the sixteenth-century Hispanic Atlantic. The genealogy of each word provides an example of how identities are built upon the disassociation from the animal (in the negation of one's connection with the animal) and on one's association with the animal (the affirmation of one's connection with the animal). The following discussion of the etymology of each term elucidates how, like in the case of *castizo,* anthropocentrism informs the meaning of each word in the sense that each term reflects the linguistic process of animalization and then humanization through the rejection of that animalization.

Mulatto

Many mixed race peoples today have chosen the term Mulatto as a term of pride in their mixed white and black heritage in reaction to the nineteenth- and twentieth century North American disparaging racial discourse centered on the Mulatto.

The term Mulatto derives from the sixteenth century Spanish Atlantic zoological thinking as it informed new modes of racial identity. Just as is the case with *perro* (dog), the word *caballo* (horse) has a long history associated with one's noble versus commoner status. In the medieval period, the horse was a significant animal in ennobling and proto-race thinking. Because of this, it would also be a significant sign in othering in the early modern period. Already the medieval tradition linked the hybrid or impure horse with impurity.[4] Following on this tradition, the useless or impure horse became a sign of the racial impurity in the emerging sixteenth century discourse of race.[5] The slander of the degenerate horse was turned into a racial stigma in the sixteenth century when the term *mulato* (little mule) entered into legal discourse in the Americas (Martinez 164–5).

Covarrubias uses the example of the horse for the entry under *castizo*. He also chooses the horse for the exemplary animal when he states purity of breed using the term *raza*, a less-used term, but roughly-equivalent to *casta*. Under the entry *raza* he gives the example of the horse as an example of *de pura raza* or purebred. In opposition to the purebred horse, Covarrubias points to the origins of *mulato* when he writes that just as the mule is the extraordinary mix of two different animal types, the horse and the donkey, so the *mulato* is the extraordinary mix between a white male and black female. Stevens carries over the same meaning in his English-Spanish dictionary when he writes under the entry for *mulato*: "mulatto, the son of a black and of a white, so call'd by reason of the mixture; from *mula*, a mule, which is a mixt breed" (Nieto Jiménez and Alvar Ezquerra 7: 6916). *Mulato*, along with other terms that indicated horse specimens, was used extensively in the New World racial configurations of eighteenth-century caste paintings.[6] In early

4 The figure of the impure horse infused oppressive discourse against religious others in the Middle Ages. David Nirenberg cites the use of *raza* for horses in medieval husbandry manuals that describe horses as connected with proto-formations of race (Nirenberg "Race"; "Was"). Fifteenth-century burlesque literature used the term *raza* to mean a defective lineage as it connected with horses, particularly in defamatory literature against the Jews. In 1407 Francisco Imperial scorns the "Judaic beast" ("bestia Juderra") and directly uses the word *raza* to mean defective Jewish lineage (Nirenberg, "Was" 249). Fifteenth-century writers connected the *alboraique*, Muhammad's fabled horse-like animal, with New Christians, claiming that those people were neither Jews nor Christians, but a kind of unnatural third species.

5 Aside from the emergence of the notion of the *mulato*, many other examples can be cited of versions of the impure horse that were used to defame people. For instance, a popular early sixteenth-century legend circulated in which the prince of the exiled Jews was called "Chamorra," a reference to a degenerate horse. This case directly connects the Jew to an animal and indirectly disparages circumcision since the term "Chamorra" refers to an ass, an infertile horse-mule offspring whose hair has been sheared (Beusterien, "The Jewish" 37–8).

6 Aside from *mulato*, eighteenth-century caste paintings include the following defamatory logic that associated the human with the horse: *albarazado* (a white-spotted horse), *barcino* (a horse or animal with colored spots), and *cambujo* (in the seventeenth century this Arabic borrowing referred to the dark toque that women wore over their faces and in the eighteenth it would refer to a reddish black horse) (Katzew 44).

modern colloquial Spanish, the impact of the slander of *mulato* was sometimes given extra pejorative emphasis—the person was *more* sub-human—through its association with the dog. The penitent Fray Martín in the seventeenth-century in the viceroyalty of Peru is reputed to have employed the horse and dog-based slander in auto-flagellation since sources state that, as he whipped himself, he called himself a "mulatto dog" (Alves 170).

Creole

Many Latin American historians use the word *criollo* to refer to a person born in America but of European origins. They often use *criollo* to describe cultures and peoples with a distinctly Latin American character. *Criollo* is generally considered a cultural, not racial category. José Juan Arrom cites sources from growing independent movements that distinguish the European from the *criollo*, infusing the word with a uniquely Latin American spirit. But *criollo* encompasses more than just an identity category in the Spanish-speaking world, especially when one considers the fact that it is the origin of the English word "Creole." Numerous studies from academic fields study Creole cultures that are not Spanish-speaking, linking the term to a rich variety of cultures, languages and peoples from across the globe.[7]

Like Mulatto, the contemporary term Creole derives from the sixteenth-century Iberian Atlantic zoological thinking as it informed new identity categories. The sixteenth century is so significant in the ennobling process because the period consolidated—as a result of Spain's American experience—the anthropocentric process into the symbolic "branding" or coining of the language of race for the first time.[8] The domesticated horse was a necessary ally, both physically and symbolically, in the successful conquering and placating of Native American populations. Also, upon establishing settlements in the Americas, Spain wanted to see its economy based on animal domestication—largely based on sheep, but also cattle and pigs—successfully transplanted on its newly conquered lands and projected the vocabulary of animal domestication and breeding into the logic of its colonial citizens, especially sexual reproduction. The meaning of the term *criollo* connected to these zoological imperial desires, especially the Spanish attempt to effective control the reproduction of the domestic animal.[9]

---

[7] Françoise Lionnet refers to Creole languages and people in Haiti, Jamaica, Papiamentu, Mauritius, Reunion, Sierra Leone, Cape Verde, Indonesia, and North America. She writes: "the common perception of Creole peoples and languages is still shrouded in ignorance and mired in exotic clichés" (28).

[8] Owning and controlling chattel and its reproduction is a gesture of ennoblement in different cultures across historical epochs. Animals and humans were literally branded in early modern Spain (Beusterien, *An Eye* 135).

[9] For a general overview of the arrival of animals in the New World, see Crosby. For the northern border of New Spain, Sandra Myres writes "Wherever the Spanish approached the frontier, horses, cattle, sheep and goats became a part of the process of imperial defense and empire building" (11).

Although *criollo* is a Portuguese-inflected diminutive form of the Spanish word *crío*, the original meaning of the word goes unnoticed by academics in the humanities and the peoples today who associate themselves with the many variations of the word. The word *criollo* was colloquial in the sixteenth century, whose meaning was originally connected to the standard Spanish *crío* and *cría*, words that connected to animal reproduction. *Crío* and *cría* could mean both the offspring of animal and the word *cría* meant the practice of raising animals. In the first modern Latin-Spanish dictionary (1492), Antonio de Nebrija writes that *cría* refers to chicks; in the first Gallician-Spanish dictionary (1536), Bachiller Olea refers to the *cría* of livestock; in his children's thesaurus (1615), Bernabé Soler refers to the *cría* of a pig; and in his French-Spanish dictionary, Antoine Oudin writes that *cría* is the same as a *cachorro* (puppy) (Nieto Jiménez and Alvar Ezquerra 4: 3071). Confirming the zoological register of *cría*, Covarrubias provides a definition that links it exclusively to animals: "Cría, raising animals like raising lambs, goats, pigs, etc. In the aviary realm, chickens, pigeons, partridges, etc. Also hares, rabbits. Fish. Raising silkworms" ("Cría, la procreación de los animales, como la cría de corderos, cabritos, lechones, etc. De las aves, como de las gallinas, palomas, perdices, ect. De liebres, Conejos, etc. De la pesca. La cría de los gusanos de la seda...") (my trans.; 369). Early definitions of *cría* or *crío,* the related noun forms of *criollo*, indicate that it did not refer to human offspring. Whether referring to chickens, pigs, rabbits or even silk works, the term *cría* was restricted to zoological reproduction. Identical in meaning to Nebrija's definiton of *cría* as "chick" or "chicken," José Juan Arrom states that the earliest usage of *criollo* associated it with the chicken. In contrast to the free range variety, *gallinha crioula* in Portuguese refers to chicks raised at home ("Criollo" 12).[10]

As a result of Spain's American experience, sixteenth-century Spanish assigned the zoologically-inflected *criollo* to human categories. In the sixteenth century, when *criollo* appears for the first time in the Spanish language, it connected to the discourse of early racial formation. Garcilaso de la Vega attributes the origin of the *criollo* to Afro-Hispanics: "It is a word that the Blacks coined" ("Es nombre que lo inventaron los negros" [607, 424]). According to Garcilaso, the word entered the Spanish language through the Portuguese-inflected dialect spoken by many Afro-Hispanics in the period. Garcilaso's assertion that the word *criollo* surfaced in the Afro-Hispanic linguistic community may be correct, but the word was quickly assimilated into legal registers that stigmatized the very blacks that

---

[10] Just as in Portuguese, modern usages of the Spanish word *criollo* refer to a wide variety of domestic animals. Generally, contemporary meanings of *criollo* refer to the many domestic animals that arrived to America from the Old World, but that were born and bred in the Americas. Hence, numerous contemporary zoological studies focus on animals whose breed is know as *criollo*, including chickens, goats, cattle and horses. These studies assume that these species are autochthonous animal breeds unique to the Americas, but whose origins are in Europe, generally Iberia. Rouse argues, for instance, that all pre-1800 cattle in the Americas were "Criollo," that is, cattle whose descendants were from an original Spanish gene pool (90).

supposedly had coined the word. In the sixteenth century the zoological register of *criollo* transferred to the legal field since sixteenth century black slaves' bills of sales in Spanish America use the word *criollo* for those who were natives to the Spanish empire (Arrom 13). The racial register of the term referred to the offspring of a bi-racial union in which one member, generally the male, was European.[11] Reflecting the meaning of *criollo* as an animal of undetermined progeny, in these early human examples *criollo* was pejoratively used, describing a person that has a partial or incomplete European connection and also suggesting that person's connection to the animal. The early modern *criollo* was animalized or othered in the sense that he or she was not completely human, having a non-Spanish parent, generally the mother. *Criollo* could also refer to an Afro-Hispanic that may come from Spain, but, because of African origins, was somehow not *castizo* or a full-fledged Spaniard.

Maroon

The term *cimarrón* is an important contemporary ennobling identity category. It is the origin of the word Maroon and in the French and Spanish Caribbean today, Maroon identity is a sign that connects to the heritage of the runaway African slaves that formed communities outside of the control of early colonial authority. A widely translated and influential novel about Esteban Montejo, a slave who becomes a freedom fighter in the wars against Spanish colonialism, is Miguel Barnet's novel, *Autobiography of a Runaway Slave* or *Biografía de un cimarrón* (1966).

Like Mulatto and Creole, the term *cimarrón* originated from zoological thinking in the Iberian imperial world. Spain's American experience from its first voyages was marked by the success and failure of transplanting animal husbandry to the Americas, that is, the need to properly breed and raise (*criar*) its animals. The etymology of *cimarrón* is connected to the futile imperial attempt to control the domestic animal, most especially, its reproduction. Since the medieval period, domestic animals were a sign of value; they were *ganado*, a word that means both earnings and herd. The semantic terms for feral, once domesticated animals that escaped as part of failed imperial expeditions profoundly impacted the formation of the vocabulary of identity in the Americas. Columbus carried horses, pigs, cattle, chickens, sheep, goats and dogs on that 1493 voyage, and, upon arrival, the animals escaped from proper breeding and husbandry procedures as practiced on the Peninsula (Crosby). From Columbus' second voyage onward, the Spanish

[11] In the seventeenth century the word *criollo* first appears in European dictionaries, such as Girolamo Vittori's *Treasure of the French, Italian, and Spanish Languages* (*Tesoro de las tres lenguas francesa, italiana y española*) (1609) and John Minsheu's *The Guide into Tongues* (1617). The seventeenth-century sources generally define *criollo* as "those that are borne of the Spaniard and Indians" (Nieto Jiménez and Alvar Ezquerra 4: 3071). By the end of seventeenth century, dictionaries define *criollo* in slightly different registers such as in Stevens 1706 "the son of a Spaniard and a West Indian woman" (Nieto Jiménez and Alvar Ezquerra 4: 3071). Also see Arrom, "Criollo."

imperial experience was marred by the trauma of *not* being able to control the domesticated animal on American soil.

The origin of the word *cimarrón* was first taken from animal vocabulary, referring variously to pigs, cattle, sheep, and dogs. In the sixteenth century the word meant, the "animal that escaped into the wild." Oviedo writes that in 1516 the pigs that went to the peaks of the mountains (*las cimas* in Spanish) were known as *cimarrones* (Arrom, "Cimarrón").[12] Other sixteenth century sources referred to renegade cattle, cats, dogs, and pigs as *cimarrones* (Arrom "Cimarrón"; Alves 87). Reflecting the old Spanish usage, in English the word "Cimarron" refers to a breed of wild sheep that could be found in the northern reaches of New Spain. (It also meant no-man's land in New Mexico and Oklahoma: today a river and a region there are named "Cimarron".) What many consider the national dog of Uruguay, the *cimarrón*, is an example of stray dogs that arrived with the Spanish that later broke off in wild bands (as Spanish dogs across the Americas), and then were domesticated once again.[13] The *cimarrón* referred to those animals that escaped the breeding control of their domestic lords by escaping to the mountains. But the meaning of "mountains" or *cimas* conflated with the meaning of any region over which the Spanish did not have control and, essentially, the *cimarrón* was then any animal that had escaped its breeder to the wild.[14]

Mustang

The Mustang became a significant sign of identity, especially of a free, independent spirit of the horse in the legend of the West as it formed North American identity.

Like *cimarrón*, the term *mesteño,* the origin of the English word "Mustang," was common in sixteenth-century documents from the Spanish Americas and, like

---

[12] The pig was an especially important animal that was used in the emerging anthropocentric language of race. On the Peninsula, medieval thinking associated the *marrano* both with the pig and those with impure Jewish blood. Also with respect to the porcine and the influence of America on race formation, the word "hybrid" had both human connotations and could refer to "half-wilds," that is, the progeny of the wild and tame pig (Hill, *Casta* 239).

[13] Wild dogs were a major worry of early chroniclers. Acosta repeats the preoccupation on two occasions in his *Natural and Moral History of the Indies* (in Book 1, Chapter 21 and Book 4, Chapter 33). He states: "Dogs have multiplied so excessively that they hunt in packs and have become so savage that they do as much harm to the herds as wolves, which is a serious problem in those islands" (65). He later repeats: "In Hispaniola dogs have multiplied both in number and size to the point that they are the plague of that island, for they eat the cattle and roam the countryside in packs. A bounty is offered to those who kill them just as it is done with wolves in Spain" (231).

[14] From a very early period, the mountain was associated with "wild regions." One source indicates that examples of the mountain as the breeding place for the "wild" animal not only included cattle, horses, and dogs, but also bees. In contrast with the domestic and tame bees that gave good honey, the non-domesticated bee, considered a wild and bastardized, was considered "mountainous" (Hill, *Casta* 238–9).

cimarrones, *mesteños* commonly referred to domesticated animals that escaped, especially stray animals like cattle and horses.[15] Taken from Spain's northern border, the Mustang, the ennobled horse of the North American Southwest, would adopt the opposite characteristics that made the horse a noble sign in Europe, that is, a breed controlled and owned by a breeder. The word, taken from "Mesta" (implying the wool produced by the Spanish Mesta economy), referred to those animals that had no owner and that the Spanish had failed to brand as its own.[16] England had employed the breeding of the Spanish horse into its own othering of Spain, but North America radically re-configured the meaning of the horse, converting the othered quality of the horse (the lack of control in its breeding) into its ennobling Mustang quality.

Mestizo

Mestizo was the most significant word in early racial formations whose earliest origins connected it to the canine. Like the word *criollo*, the word *mestizo* was highly influential in nineteenth-century Latin American national and transnational identity formation. The figure of the *mestizo* was a sounding board for the foundational myth of Spanish American nationhood and identity and the language of mixture or mongrelization (mestizoness) became the banner for Latin American identities such as in the transnational notion of a Latin American in José Vasconcelos' *Cosmic Race*. A unique American *mestizo* identity, like *criollo*, also informs academic projects in Hispanism such as Estelle Taricas' *The Inner Life of Mestizo Nationalism*.

The term *mestizo* originally surfaced from zoological registers, specifically the context of canine husbandry. In his description of the Spanish mastiff, Gabriel Alonso de Herrera describes a *mestizo* as a mix between a hunting and shepherding dog (348; Hill, "Casta" 236). The canine term transfers in the mid-sixteenth century to the human realm in the American context. Already by 1540, *mestizo* transferred to the sphere of human categorization and it had a racial-religious connotation, meaning the offspring mix between a black Moor and a

---

[15] A Disney movie, *Spirit: Stallion of the Cimarron* combines two words of similar etymological origin, the wild place "Cimarron" with the horse "Spirit" who is a mustang.

[16] The Mesta was a stockman's association or a group of animal owners (most especially sheep) and in the Middle Ages the animals that strayed were known as *mesteños*. Julius Klein explains that occasionally strays from the sheep migrations "were called mestas, though this was not common; they were usually designated as *mesteños* or as *mostrencos*, the general term applied to all ownerless property" (10). The Mesta's primary concern was the assignment of stray animals, both migratory and sedentary to their rightful owners and with the sale of all unclaimed strays (*mesteños*). In 1778 Teodoro de Croix, Commandant General of the Interior Provinces, declares all wild cattle and horses which he calls *mesteños* to be property of the Spanish Crown (Myres 26). For more on Mesta, see Alves 72–85.

Christian (Martínez 164). As European dictionaries confirm, the American use of the word transferred back to Europe and, as the seventeenth century advanced, *mestizo* was a general classifier for mixed peoples. Generally, *mulato* came to refer to the mix of a black and white Spaniard, while *mestizo* to the mix of Indian and Spaniard. Lorenzo Franciosini in his Italian-Spanish dictionary (1620) calls the *mestizo* "a bastard" and Pedro de Salas in his *Thesaurus hispanolatinus* (1645) calls it a "hybrid" (Nieto Jiménez and Alvar Ezquerra 7: 6686). John Minsheu, in his Spanish-English dictionary (1599) defines *mestizo*: "that which is come or sprung of a mixture of two kinds, as a black-Moore and a Christian, a mungrell dog or beast" (Nieto Jiménez and Alvar Ezquerra 7: 6686).

Cholo

In modern usage in the United States, Cholo often refers to a Mexican or Mexican American. Since the 1960s in North America, Cholo has been a term of pride among some Hispanic groups. The identitarian meaning of Cholo emerged among the subcultures of barrio gangs from southern California. One of the earliest semantic ennobling moves of Cholo occurred in the eighteenth century. In the Andean region *cholo* refers today to mixed race peoples, but people—in the eighteenth century—referred to María Michaela Villegas Hurtado as "Perricholi," a word that roughly translates as "Bitch Dog." The nickname, however, was not disparaging, but became a sounding of pride in Perricholi's renowned ability as an actress in Golden Age Spanish plays and Perricholi later became an iconic figure of Spanish-speaking American culture, featured as the subject of ballets, novels and films.

The word *cholo* simply meant "dog" in many Mesoamerican languages and it is a semantic cousin of the Xolo, the contemporary dog breed.[17] Cholo first entered into the Spanish language shortly after Spain's colonial experience in the Americas. When the American word "xolo," first transferred to the Spanish language, like *mulato*, it was often charged with negative racial connotations in the early Hispanic world. *Cholo*'s meaning as dog rarely appears in colonial texts.

---

[17] *Cholo* (or Xolo as it is called today in English) relates to Xolotl, the Aztec canine deity, and specifically, it comes from *xoloitzcuintli* which literally means the *itzcuintli*, the Nahuatl term for dog, that belongs to Xolotl. The Xolo is a hairless dog, known in the Greater Antilles, Paraguay, Peru and Argentina, but most famously found in ancient and modern-day Mexico where they held and hold special status. Xolos were bred for consumption and human companionship under the Aztecs. Bernardino de Sahagún is one of the few sixteenth-century sources to mention their indigenous names when he writes: "They raised other little dogs called *xoloitzcuintli* that hardly had any hair and at night they would wrap them up in blankets" ("Otros perrillos criaban que llamaban xoloitzcuintli que apenitas ningún pelo tenían, y de noche abrigábanlos con mantas para dormir" [my trans.; qtd. in Czitrom 17]). Connected to the deity, the *xoloitzcuintli* was extremely important in negotiating the transitional place between death and life. Xolotl carried the sun through the land of the dead and is represented with a dog's head or as an entire dog.

Rather, it immediately transferred to the logic of human racial groups. A *cholo* could be a servant or a slave and was used in caste paintings to describe a mulatto-black offspring. Guaman Poma de Ayala writes that *cholos* are the children of *mulatos* who he defines as the offspring of a black and an Indian woman or an Indian and a black woman. Garcilaso Inca de la Vega also defines the *cholo* as the offspring of *mulatos*. But Garcilaso Inca de la Vega also recognizes that word originally meant dog when writes that a *cholo* is a mutt or non-purebred. Garcilaso equates the *cholo* with the common *gozque*, the canine variety that the Spanish disparaged as feminine and useless (Chapter 4). Aside from Mestizo and Cholo, other canine-inflected words, such as *grifo*[18] and *zambo*[19] would also be used in newly-constructed racial discourse and caste paintings.

---

18 In the early modern Spanish period, the hybrid lion-eagle mythological griffin or *grifo* often appeared in literature. Covarrubias calls the *grifo* a monstrous animal with head and beak of eagle, vulture's wings, lion's body and serpent's tail. It most famously appeared in Calderón de la Barca's *La vida es sueño* not as griffin, but as *hipógrifo*, the griffin's manifestation that includes a horse. While *grifo* was associated with the griffin, the hybrid creature from classical mythology, *grifo* would form part of the emerging discourse of race formations. The original root of the word was a Greek word meaning "twisted" or "curled." Certain birds brought back from the Americas were called griffins since *grifo* and *grifón*, from Late Latin *gryphus*, denoted a curved or hooked nose, becoming a metonymy for a bird with a hooked beak. Seventeenth-century natural historians like Juan Eusebio Nieremberg explained the arrival of griffins when he writes that the mythological creature did not exist and that the griffin was simply a large and fierce fowl (like a condor) (French and de Asúa 45). The large, monstrous quality of the medieval griffin became a name for a dog in the Middle Ages (Owst) and by the early modern period it became a type of dog. Interestingly, the griffin became a small dog variety in the early modern period. Griffin (or *grifo* in Spanish) meant a "dog," particularly curly-haired (a connection was made with "curved"), small, hybrid, and mutt-like dog varieties. Today, various types of Griffons are classified as dog breeds and *grifo* in Spanish was a small curly-haired variety and painters often represented it with women. In caste paintings it referred to those of African descent with curly hair such as in one caste painting by an unknown artist, *From Mulatto and Albino, Grifo* (ca. 1770) (Katzew 152). When caste paintings use the word *grifo* they conflate the semantic connotations of hybridity, the animal, and the canine.

19 In the twentieth century, the Sambo figure would stir up black–white racial tensions in the twentieth-century United States. As many Spanish words transferred into the English language (such as "Negro" and "nigger" from the Spanish *negro*), so Sambo was a semantic descendant of *zambo*. The Spanish word *zambo*, meaning knock-kneed or bowlegged, was associated with the physical features of Africans and Afro-Hispanics in sixteenth-century Spain and it also meant a monkey or, like the *grifo*, it was a hybrid creature and sometimes characterized as a bushy-haired dog. In the eighteenth century *zambo* (or its common variation *zambaigo*, "son of zambo") was often found in caste paintings to mean the offspring of an Indian and a black (Katzew 44). In his English-Spanish dictionary, under the entry *zambo*, Stevens mentions its meaning as a physical attribute and as a child of a mulatta and Indian: "splay-footed; sambo, the son of a mulata woman and an Indian man" (Nieto Jiménez and Alvar Ezquerra 10: 10,034).

Aside from the connection between animals and the emergence of race, further study is needed on the connection between the animal and nationalism. In the late medieval period, the word for dog in Spanish was *can*, a word that interchanged with the more frequent *perro*. The word *gozque* in early modern Spain served as a substitute for *perro*, indicating the pejorative meaning of the dog in the popular imagination—dogs served no practical function. In parallel to *gozque*, the word *rocín* (remember Quijote's horse Rocinante) was considered a useless horse. *Equus* meant horse in classical Latin. In classical Latin, the word *caballus* (which would give the Spanish word for horse *caballo*) meant a nag or useless horse. In Spanish, *caballo* became what in Latin was *equus* and *rocín* became what was in Latin *caballus*. The same semantic evolution occurred with respect to the Spanish word for dog. *Perro* replaced *can* like *caballo* replaced *equus*. In turn, the semantic emergence (but loss) of the word *gozque* can be compared to emergence (but loss) of *rocín*. The *gozque* and the *rocín* were not considered a true dog and horse because each was not a useful animal. The words that implied the lesser horse (*rocín*) and lesser dog (*gozque*) became lexical items in the sixteenth century, the period that corresponded with the flourishing humanist interest in words and their origin. Each word, however, disappears after the nineteenth century, indicating the linguistic aversion in Spanish toward the concept of an "impure" animal. This aversion was especially acute in the eighteenth and nineteenth century which inspires the following questions: why did common words that implied zoological impurity like "mongrel" disappear from Peninsular Spanish? To what degree does the figure of the animal in the collective mindset play a role in the definition of the modern sense of the Spanish nation?[20]

[20] In contrast to Latin America, the word for a mixed breed dog would be eliminated from Peninsular Spanish. Today, the word for "mongrel" in Colombia for instance is still *gozque*, but Peninsular Spanish lost the word *gozque* and has no standard word for "mongrel." Many in Spain use *perro cruzado* (crossed dog), but no standard word exists for mongrel. In some regions in Spain *chucho* is used and local expressions such as *mil leches* (suckled from a thousand) or *perro de cuatro esquinas* (the dog from four corners). Although *gozque* is preserved in Colombia, other lexical variations exist for mongrel such as *sato* in Puerto Rico, *quiltro* in Chile, and *pichicho* in Argentina. The lexical vacuum—the lack of a word for mutt in Peninsular Spanish—reflects the erasure of the notion of mestizoness as Spain construed its national identity in the eighteenth and nineteenth century. Interestingly, the word "mutt" does not appear in North American English until the twentieth century. The US portrayed itself as a melting pot, not a *mestizo* place of difference, and, the lack of the word "mutt" reflects a parallel case of ideological denial as in Spain: in contrast to the Spanish Americas, the construction of nineteenth-century national identity in the cases of Spain and North America depended on a mongrel-free homogenized racial identity.

Works Cited

'Abd Allah Ibn 'Umar al-Bayzar, Muhammed Ibn (Moamín). *Libro de los animales que cazan (Kitab al-Yawarih)*. Ed. José Manuel Fradejas Rueda. Madrid: 1987. Print.

Acosta, José de. *Historia natural y moral de las indias*. Ed. José Alcina Franch. Madrid: Dastin, 2002. Print.

———. *Natural and Moral History of the Indies*. Ed. Jane E. Mangan. Intro. Walter D. Mignolo. Trans. Frances M. López-Morillas. Durham: Duke University Press, 2002. Print.

Agamben, Giorgio. *Homo sacer: Sovereign Power and Bare Life*. Trans. Daniel Heller-Roazen. Stanford: Stanford University Press, 1998. Print.

———. *The Open: Man and Animal*. Trans. Kevin Attell. Stanford: Stanford University Press, 2004. Print.

Alcázar, Baltasar del. *Obra poética*. Ed. Valentín Núñez Rivera. Madrid: Cátedra, 2001. Print.

Alemán, Mateo. *Guzmán de Alfarache*. Ed. Francisco Rico. Barcelona: Planeta, 1983. Print.

Alfonso XI. *Libro de la montería*. Ed. María Isabel Montoya Ramírez. Granada: University of Granada, 1992. Print.

"El alano:" poema anónimo del siglo XVII. Sevilla: E. Barco, 1902. Print.

Alonso, Carlos J. "Spanish: The Foreign National Language." *Profession 2007* 144 (2007): 55–60. Print.

Altozano, Manuel. "La élite morisca de los boneteros." *El País*. 29 Jan. 2005. Web. 11. Dec. 2008.

Alvárez López, Enrique. "El perro mudo americano." *Boletín de la sociedad española de historia natural* 40 (1942): 411–17. Print.

Alves, Abel A. *The Animals of Spain: An Introduction to Imperial Perceptions and Human Interaction with Other Animals, 1492–1826*. Leiden: Brill, 2011. Print.

Amezúa y Mayo, Agustín G. *Cervantes: Creador de la novela corta española*. 2 vols. Madrid: CSIC, 1965. Print.

Argote de Molina, Gonzalo. *Discurso sobre la montería*. Madrid, 1882. Print.

Aristotle. *History of Animals, Books VII-X*. Ed. and trans. D.M. Balme. Cambridge: Harvard University Press, 1991. Print.

Arrom, José. "Cimarrón: apuntes sobre sus primeras documentaciones y su probable origen." *Revista española de antropología americana* 13 (1983): 47–57. Print.

———. "Criollo: Definición y matices de un concepto." *Certidumbre de América: Estudios de letras, folklore y cultura*. Madrid: Grelos, 1971. 11–26. Print.

Asúa de, Miguel and Roger French. *A New World of Animals: Early Modern Europeans on the Creatures of Iberian America*. Hants: Ashgate, 2005. Print.

Baillie-Grohman, W.A. "The Finest Hunting Manuscript Extant." *The Burlington Magazine for Connoisseurs*, 2.4 (1903): 8–21. Print.

Baker, Steve. *Picturing the Beast: Animals, Identity and Representation.* Manchester: Manchester University Press, 1993. Print.

Barahona de Soto, Luis. *Diálogos de la Montería*. Ed. José Lara Garrido, Malaga: ADR-NORORMA, 2002. Print.

Bass, Rick. *Brown Dog of the Yaak: Essays on Art and Activism*. Minneapolis: Milkweed, 1999. Print.

Beilin, Katarzyna. "Men with Animal Heads: A Revised 'Humanity' in the Theater of Juan Mayorga and Spanish Debates on Culture." *Academia.edu*. Web. 31 May 2012.

Benavente, Jacinto. "Nuevo coloquio de los perros," *El cuento semanal*, 93 (9 October 1908): n.p. Print.

Benavente, Luis Quiñones de. "El sueño del perro." *Teatro breve del siglo de oro*. Web. 2 Nov. 2009.

Benito Ruiz, Eduardo de and Rafael Malo Alcrudo. *Las razas caninas españolas.* Barcelona: Editorial Hispano Europa, 1992. Print.

Beusterien, John. "Did Cervantes Stutter?" *Cervantes: Bulletin of the Cervantes Society of America* 29.1 (2009): 209–20. Print.

———. *An Eye on Race: Perspectives from Theater in Imperial Spain.* Lewisburg: Bucknell University Press, 2006. Print.

———. "Goya's Dogs." *Teaching the Transatlantic Eighteenth Century.* Eds. Jennifer Frangos and Cristobal Silva. Newcastle upon Tyne: Cambridge Scholars, 2010. 155–72. Print

———. "The Spanish 'Jewish Conspiracy' Letters," *Monographic Review* 22 (2006): 33–45. Print.

Beusterien, John and J. Baird Callicott. "Humor and Politics through the Animal in Cervantes and Leopold." *Comparative Literature Studies* (Penn State University Press), forthcoming.

Bloom, Harold. *Shakespeare: The Invention of the Human*. New York: Riverhead Books, 1998. Print.

Boehrer, Bruce, ed. *A Cultural Dictionary of Animals in the Renaissance*. Oxford: Berg, 2007. Print.

Boehrer, Bruce. *Animal Characters. Nonhuman Beings in Early Modern Literature.* Philadelphia: University of Pennsylvania Press, 2010. Print.

———. "Animal Studies and the Deconstruction of Character." *PMLA: Publications of the Modern Language Association of America* 124 (2009): 542–47. Print.

———. *Parrot Culture: Our 2,500-Year-Long Fascination with the World's Most Talkative Bird.* Philadelphia: University of Pennsylvania Press, 2004. Print.

———. "Shylock and the Rise of the Household Pet: Thinking Social Exclusion in The Merchant of Venice." *Shakespeare Quarterly* 50 (1999): 152–70. Print.

Booth, Roy. "Early Modern Pets." *Early Modern Whale*. 13 May 2006. Web. 12. Dec. 2008.

Borja, Juan de. *Empresas morales*. Ed. Bravo-Villasante, Carmen. Madrid: Fundación Universitaria Española, 1981. Print.

Borrego Gutiérrez, Esther. "Canciones populares en el teatro cómico breve del siglo XVII." *Los géneros breves de la literatura tradicional. Actas del congreso internacional celebrado en la Universidad de Alcalá, 28–30 octubre 1998*. Ed. Carlos Alvar Ezquerra. Alcalá de Henares: University of Alcalá, 2001. 179–88. Print.

Bowra, C.M. *Early Greek Elegists*. London: Oxford University Press, 1938. Print.

Bowron, Edgar Peters, Carolyn Rose Rebbert, Robert Rosenblum and William Secord. *Best in Show: The Dog in Art from the Renaissance to Today*. New Haven: Yale University Press, 2006. Print.

Broadfoot, Keith. "Las Meninas and the King's Two Bodies." *Word & Image* 17.3 (2001): 219–32. Print.

Brown, Jonathan. *Las Meninas: Imágenes e ideas en la pintura española del siglo XVII*. Madrid: Alianza, 1985. Print.

———. *Velázquez*. New Haven: Yale University Press, 1986. Print.

Buero Vallejo, Antonio. *Las Meninas*. Ed. Juan Rodríguez Castellano. New York: Charles Scribner's Sons, 1963. Print.

Bugnion, Jacques. *Les chasses médiévales: Le brachet, le lévrier, l'epagneul, leur nomenclature, leur métier, leur typologie*. Gollion: Infolio, 2005. Print.

Caesar, Terry. *Speaking of Animals. Essays on Dogs and Others*. Leiden: Brill, 2009. Print.

Caius, Johannes. *Of Englishe Dogges*. London: Da Capo Press, 1576. Print.

Calderón de la Barca, Pedro. "La rabia." *Biblioteca Virtual Miguel de Cervantes*. Eds. Evangelina Rodríguez and Antonio Tordera. Web. 2 Nov. 2009.

Cantó, Josefa, ed. Plinio. *Historia Natural*. Madrid: Catedra, 2002. Print.

Cárdenas, Anthony. "Berganza: Cervantes's Can[is] Domini." *Cervantes and the Pastoral*. Eds. José J. Labrador and Juan Fernández Jiménez. Cleveland: Penn State University and Cleveland State University, 1986. 19–35. Print.

Carranza, Paul. "Cipión, Berganza, and the Aesopic Tradition." *Cervantes: Bulletin of the Cervantes Society of America* 23.1 (2003): 141–63. Print.

Castellanos, Juan de. *Elegías de varones ilustres de indias. Primera parte. Obras de Juan de Castellano*. Bogatá: Editorial ABC, 1955. Print.

Castro, Américo. *The Spaniards: An Introduction to Their History*. Trans. Willard F. King and Selma Margaretten. Berkeley: University of California Press, 1971. Print.

Cavalieri, Paola. *The Death of the Animal*. New York: Columbia University Press, 2009. Print.

Cervantes, Miguel de. *The Adventures of Don Quixote de la Mancha*. Trans. Peter Anthony Motteux. New York: A.A. Knopf, 1939. Print.

———. *The Dialogue of the Dogs*. Trans. David Kipen. New York: Melville, 2008. Print.

———. *The Exemplary Novels of Cervantes*. Middlesex: Echo Library, 2007. Print.

———. *Don Quijote de la Mancha.* Ed. Francisco Rico. Barcelona: Crítica, 2001. Print.

———. *Don Quixote.* Ed. Annette Grant Cash & Victoria Richardson. Trans. Tom Lathrop. Newark, Delaware: Cervantes & Co., 2005. Print.

———. *Don Quixote.* Trans. Edith Grossman. New York: Harper Collins, 2003. Print.

———. *Don Quixote.* Trans. Tobias Smollet. New York: Oxford University Press, 1992. Print.

———. *Don Quixote de la Mancha.* Ed. Francisco Rico. Trans. John Rutherford. New York: Penguin Books, 2000. Print.

———. *Don Quixote: The Ingenious Gentleman of La Mancha.* Trans. John Ormsby. Norwalk, Conn: Heritage Press, 1984. Print.

———. *Don Quixote.* Trans. Watts, Henry Edwards. London, 1887–89. 5 vols. New York: Appleton, 1898. Print.

———. *Don Quixote de la Mancha.* Trans. Ed. Charles Jarvis. Oxford: Oxford University Press, 1992. Print.

———. *The History of Don Quixote of the Mancha.* Ed. W.E. Henley. Trans. Thomas Shelton. London: David Nutt, 1896. Print.

———. *The History of That Ingenious Gentleman Don Quijote de La Mancha.* Trans. Burton Raffel. New York: Norton, 1995. Print.

———. *The Ingenious Gentleman Don Quixote de la Mancha.* Trans. Samuel Putnam. New York: Viking, 1949. Print.

———. *Novelas ejemplares,* vol. II. Ed. Harry Sieber. Madrid: Cátedra, 1989. Print.

Chaucer, Geoffrey. "The Wife of Bath's Tale." *The Canterbury Tales.* Web. 20 Nov. 2008.

"Chihuahua." Mimi.hu. Web. 20 Oct. 2009.

Checa, Fernando. *El coleccionismo en España: De la cámara de maravillas a la galería de pinturas*. Madrid: Cátedra, 1985. Print.

Childers, William. *Transnational Cervantes.* Toronto: University of Toronto Press, 2006. Print.

Chiong Rivero, Horacio. "El Caballero de los Leones: la carnavalesca farsa felina en el Quijote." *Bulletin of Spanish Studies* 84.8 (2007): 1007–1027. Print.

Cid, Jesús Antonio. "Calderón y el romance de El bonetero de la Trapería." *Hispanic Review* 45.4 (1977): 421–34. Print.

Cohen, Simona. "Changing Functions of the Canine Image in Venetian Religious Paintings of the Sixteenth Century." *Ikon* 2.2 (2009): 277–86. Print.

Coile, Caroline. *The Cocker Spaniel Handbook.* Hauppauge: Barron's Educational Series, 2007. Print.

Colón, Cristobal. "Carta de Diego Alvárez Chanca." *Los documentos de Puerto Rico.* Ed. Luis M. Iriarte. 1898. Web. 11. Dec. 2008.

Corominas, Joan. *Diccionario crítico etimológico castellano e hispánico*. Madrid: Gredos, 1980–91. Print.

Correas, Gonzalo. *Vocabulario de refranes y frases proverbiales.* Eds. Louis Combet, Robert Jammes, and Maïte Mir-Andreu. Madrid: Castalia, 2000 (1627). Print.

Crosby, Alfred W. *The Columbian Exchange: Biological and Cultural Consequences of 1492.* Westport: Greenwood Press, 1972. Print.

Cruz, Anne J. "Golden Age Studies in the 21st Century: A View of the Culture Wars." In Martín-Estudillo, 81–86.

Cuadra Blanco, Juan Rafael de la. "Interludio: El perro Salomón." *El Escorial y el Templo de Salomón.* Web. 12. Dec. 2008.

Darnton, Robert. *The Great Cat Massacre.* New York: Basic Books, 1984. Print.

Derr, Mark. *A Dog's History of America.* New York: North Point Press, 2004. Print.

Delgado Díaz, Arturo. *El mastín español.* Madrid: Fuenlabrada, 1991. Print.

Delmar, Fernando. "Discurso filosófico sobre el lenguaje de los animales." *Conquista y contraconquista, la escritura del Nuevo Mundo (Actas del XXVIII Congreso del Instituto Internacional de Literatura Iberoamericana).* Eds. Julio Ortega y José Amor y Vázquez. México: El Colegio de México y Brown University, 1994. 419–26.

Derrida, Jacques. "The Animal That Therefore I Am (More to Follow)." Trans. David Wills. *Critical Inquiry* 28 (2002): 369–418. Print.

———. "'Eating Well,' or The Calculation of the Subject: An Interview with Jacques Derrida." *Who Comes After the Subject?* Eds. Eduardo Cadava, Peter Connor, and Jean-Luc Nancy. New York: Routledge, 1991, 96–119. Print.

———. *The Gift of Death.* Trans. David Wills. Chicago and London: University of Chicago Press, 1995. Print.

———. *Politics of Friendship.* Translated by George Collins. London: Verso, 1997. Print.

———. "And Say the Animal Responded?" Trans. David Wills. In Wolfe, *Zoontologies* 121–46. Print.

Diálogos de la Montería. Ed. Francisco R. de Uhagón. Madrid: La Sociedad de Bibliófilos Españoles, 1890. Print.

Diccionario de la lengua castellana. Vol. 4. Real Academia Española, 1734. Print.

Diel, Lori Boornazian. "The Spectacle of Death in Early Colonial New Spain in the Manuscrito del aperreamiento." *Death: The Case of the Early Modern Hispanic World.* Eds. John Beusterien and Constance Cortez. Hispanic Issues Online. University of Minnesota Press. 2010. 144–63. Web. 22 Dec. 2011.

Dopico Black, Georgina. "The Ban and The Bull: Cultural Studies, Animal Studies, and Spain." *Journal of Spanish Cultural Studies* 11 (2010): 235–49. Print.

Duncan, David Douglass. *Picasso and Lump: A Dachshund's Odyssey.* New York: Bulfinch Press, 2006. Print.

"El escudo de Canarias: Evolución histórica." Hispanismo.org. Web. 11 Jan. 2012.

Ellsworth Hamann, Byron. "The Mirrors of Las Meninas: Cochineal, Silver, and Clay." *Art Bulletin* 92 (2010): 6–35. Print.

El Saffar, Ruth. *Beyond Fiction: The Recovery of the Feminine in the Novels of Cervantes.* Berkeley: University of California Press, 1984. Print.

Eliav-Feldon, Miriam, Benjamin Isaac, and Joseph Ziegler, eds. *The Origins of Racism in the West.* Cambridge: Cambridge University Press, 2009. Print.

Emmens, J. A. "Las meninas de Velázquez: espejo de príncipes para Felipe IV." Marías 43–66. Print.

Espéculo de los legos. Ed. José María Mohedano Hernández. Madrid: Consejo Superior de Investigaciones Científicas, 1951. Print.

"Enxiemplo del *alano* que llevava la pieça de carne en la boca." *La transformación y adaptación de la tradición esópica en el* Libro de buen amor. Web. 11 Jan. 2012.

Epps, Brad and Luis Fernández Cifuentes, eds. *Spain Beyond Spain: Modernity, Literary History, and National Identity*. Lewisburg: Bucknell University Press, 2005. Print.

Erbez, José M. "Historia del escudo de Canarias, Siglo XVIII." *Símbolos de Canarias: Banderas y escudos de las islas.* Web. 11 Jan. 2012.

Estrin, Elana. "From the Galleries: Tycho Brahe's *Astronomiae instauratae mechanica*." Blogs: Cultural Compass. 3 Dec. 2009. Web. 15 Dec. 2011.

Evans, E. P. *The Criminal Prosecution and Capital Punishment of Animals: The Lost History of Europe's Animal Trials*. 1906. Repr. London: Faber and Faber, 1987. Print.

Fernandez, Amy. *Chihuahuas: Barron's Dog Bibles.* Hauppauge: Barron's, 2009. Print.

Few, Martha and Zeb Tortorici, eds. *Centering Animals: Writing Animals into Latin American History*. Durham: Duke University Press, forthcoming.

Fitcher, William L. "Lope de Vega an Imitator of Quevedo?" *Modern Philology* 30.2 (1932): 141–46. Print.

Forcione, Alban K. *Cervantes and the Mystery of Lawlessness: A Study of "El casamiento engañoso" y "El coloquio de los perros."* Princeton: Princeton University Press, 1984. Print.

———. *Majesty and Humanity: Kings and their Doubles in the Political Drama of the Spanish Golden Age*. New Haven: Yale University Press, 2009. Print.

Freud, Sigmund. *Civilization and its Discontents.* Scribd.com. 1930a. Web. 31 May 2012.

Fuchs, Barbara. "Golden Ages and Golden Hinds; or, Periodizing Spain and England." *PMLA: Publications of the Modern Language Association of America* 127.2 (2012): 321–27. Print.

Fudge, Erica. *Brutal Reasoning: Animals, Rationality, and Humanity in Early Modern England.* Ithaca and London: Cornell University Press, 2006. Print.

———. "The Dog, the Home and the Human, and the Ancestry of Derrida's Cat." *Oxford Literary Review* 29 (2007): 37–54. Print.

———. "'The dog is himself': Humans, Animals, and Self-Control in The Two Gentlemen of Verona." *How to Do Things with Shakespeare: New Approaches, New Essays*. Ed. Laurie Maguire. Malden, MA: Blackwell, 2008. 185–209. Print.

———. *Perceiving Animals: Humans and Beasts in Early Modern English Culture*. New York: St. Martin's, 2000. Print.

Gabilondo, Joseba. "The Impossibility of Iberian Studies: A (Trans)Atlantic Approach to 'Greater Spanish Studies' and the Excess of Iberian Nationalisms." Web. 1 June 2012.

Gale, M. "From Breast to Beard (Story of Renaissance)." *Aiwas.net.* Web. 11 Jan. 2011.

Garber, Marjorie. *Dog Love*. New York: Simon and Schuster, 1996. Print.

———. *Profiling Shakespeare*. New York: Routledge, 2008. Print.

Garcés, María Antonia. *Cervantes in Algiers. A Captive's Tale*. Nashville: Vanderbilt University Press, 2002. Print.

Garcilaso de la Vega, Inca. *Comentarios reales*. Ed. Jose de la Riva-Agüero, México D. F.: Porrúa, 1990. Print.

———. *Royal Commentaries of the Incas and General History of Peru*. 2 vols. Trans. Harold V. Livermore. Austin: University of Texas Press, 1970. Print.

George, Wilma. *Animals and Maps.* Berkeley: University of California Press, 1969. Print.

Gerbi, Antonello. *Nature in the New World: From Christopher Columbus to Gonzalo Fernández de Oviedo*. Trans. Jeremy Moyle. Pittsburgh: University of Pittsburgh Press, 1985. Print.

Gillet, Joseph E. "The 'Coplas Del Perro De Alba'." *Modern Philology*, 23.4 (1926): 417–44. Print.

Gittes, Tobias Foster. "Cañizares's Textual Auto-da-fé: Inquisitorial Dogs and Martyred Witches in Cervantes's 'Novela y coloquio que pasó entre Cipión y Berganza.'" *Hispanic Review* 74.4 (2006): 355–77. Print.

Glen, Thomas L. "Should Sleeping Dogs Lie? Once Again, Las Meninas and the Mise-en-Scène." *Source: Notes in the History of Art* 12 (1993): 30–36. Print.

Gottfried, Maya. *Good Dog.* Illus. Robert Rahway Zakanitch. New York: Knopf, 2005. Print.

Greer, Margaret R. "Diana, Venus and Borrowed Dogs: On Hunting in *Don Quixote*." *Cervantes y su mundo*. Eds. A. Robert Lauer & Kurt Reichenberger. Kassel: Reichenberger, (2005): 201–22. Print.

———. "María de Zayas: The Said and the Unsaid." *Laberinto*. 2002. Web. 13. Dec. 2008.

———. *The Play of Power: Mythological Court Dramas of Calderón de la Barca.* Princeton: Princeton University Press, 1991. Print.

———. "Mine and Thine: The Spanish 'Golden Age' and Early Modern Studies." *PMLA: Publications of the Modern Language Association of America* 126. 1 (2011): 1–15. Print.

Graf, E.C. *Cervantes and Modernity. Four Essays on Don Quijote.* Lewisburg: Bucknell University Press, 2007. Print.

Guzauskyte, Evelina. "Fragmented Borders, Fallen Men, Bestial Women: Violence in the *Casta* Paintings of Eighteenth-century New Spain." *Bulletin of Spanish Studies* 86 (2009): 175–204. Print.

Hancock, David. *The Heritage of the Dog*. Manchester: Nimrod Press Limited, 1990. Print.

Haraway, Donna. *The Companion Species Manifesto: Dogs, People and Significant Otherness*. Chicago: Prickly Pear Press, 2003. Print.

———. *When Species Meet*. Minneapolis: University of Minnesota Press, 2007. Print.

Harrison, Robert Pogue. *The Dominion of the Dead*. Chicago: University of Chicago Press, 2003. Print.

Hart, Thomas. "La novela y el romance en el Quijote," *Insula* 538 (1991): 21–23. Print.

———. "Renaissance Dialogue into Novel: Cervantes's *Coloquio*." *Modern Language Notes* 105 (1990): 191–202. Print.

Hauser, Marc D., Noam Chomsky, and W. Tecumseh Fitch. "The Faculty of Language: What Is It, Who Has It, and How Did It Evolve?" *Science* 298 (2002): 1569–79. Print.

Herrera, Gabriel Alonso. *Agricultura General*. Ed. Eloy Terrón. Madrid: Ministerio de Agricultura, Pesca y Alimenación, 1988. Print.

Hill, Ruth. "Casta as Culture and the *Sociedad de castas* as Literature." *Interpreting Colonialism*. Eds. Byon R. Wells and Philip Stewart. Oxford: Voltaire Foundation, 2004. 231–59. Print.

Höltgen, Karl Josef. "Clever Dogs and Nimble Spaniels: on the Iconography of Logic, Invention, and Imagination." *Explorations in Renaissance Culture* 24 (1998): 1–36. Print.

The Huarochirí Manuscript: A Testament of Ancient and Colonial Andean Religion. Trans. Frank Salomon and George L. Urioste. Ed. Frank Salomon. Transcription by George L. Urioste. Austin: University of Texas Press, 1991.

Huarte de San Juan, Juan. *Examen de ingenios para las ciencias*. Ed. Guillermo Serés. Madrid: Cátedra, 1989. Print.

Huggan, Graham and Helen Tiffin. *Postcolonial Ecocriticism. Literature, Animals, Environment*. London: Routledge, 2010. Print.

Jackson, James L. "Shakespeare's Dog-and-Sugar Imagery and the Friendship Tradition." *Shakespeare Quarterly* 1.4 (1950): 260–63. Print.

Johns, Catherine. *Dogs: History, Myth, Art*. Cambridge: Harvard University Press, 2008. Print.

Johnson, Sara E., "'You Should Give them Blacks to Eat': Waging Inter-American Wars of Torture and Terror." *American Quarterly* 61.1 (2009): 65–92. Print.

Kalof, Linda. *Looking at Animals in Human History*. London: Reaktion Books, 2007. Print.

Katzew, Ilona. *Casta Painting: Images of Race in Eighteenth-Century Mexico*. New Haven: Yale University Press, 2004. Print.

Keber, Eloise Quiñones. "Xolotl: Dogs, Death, and Deities in Aztec Myth." *Latin American Indian Literatures Journal* 8 (1992): 229–39.

Kemmerer, Rachel D. "A Brief History of the Papillon." *Papillon Club*. Web. 11. Dec. 2008.

Kitchell, Kenneth F. Jr. "Man's Best Friend? The Changing Role of the Dog in Greek Society." *Man and Animal in Antiquity. Proceedings of the Conference at the Swedish Institute in Rome*. Rome, 2004. Web. 23 August 2011.

Klein, Julius. *The Mesta: A Study in Spanish Economic History, 1273–1836*. Port Washington: Kennikat, 1964. Print.

Kohanski, Alexander S. *An Analytical Interpretation of Martin Buber's* I and Thou *with a Biographical Introduction and Glossary*. Woodbury: Barron's Educational Series, 1975. Print.

Lerner, Jesse. "No Hair of the Dog." *Cabinet* 6 (2002). Web. 20 Nov. 2008.

Lestel, Dominque. *The Friend of My Friends*. Trans. Jeffrey Bussolini. (*Les Amis de mes Amis* Paris: Seuil, 2007), forthcoming.

Levinas, Emmanuel. "The Name of a Dog, or Natural Rights." *Animal Philosophy: Essential Readings in Continental Thought*. Eds. Matthew Calarco and Peter Atterton. New York: *Continuum*, 2004. 47–61. Print.

Licht, Fred. "Goya's Portrait of the Royal Family." *The Art Bulletin* 49 (1967): 127–8. Print.

Lionnet, Françoise. "Cosmopolitan or Creole Lives? Globalized Ocenas and Insular Identities." *Profession* (2011): 23–43. Print.

Loa para la Comedia del sueño del perro / compuesta por un ciego de la Estafeta. Traducida en castellano y portugués por un aremenio de la Puerta del Sol, 1710. Print.

MacInnes, Ian. "Mastiffs and Spaniels: Gender and Nation in the English Dog." *Textual Practice* 17.1 (2003): 21–40. Print.

Manning, Patricia W. "Present Dogs, Absent Witches: Illustration and Interpretation of 'El coloquio de los perros'" *Cervantes* 27.2 (2007): 125–30. Print.

Marías, Fernando. "El genero de las meninas: los servicios de la familia." Marías, 247–78.

———. ed. *Otras meninas*. Madrid: Siruela, 1995. Print.

Martin, Adrienne L. "Cervantes y la canífilia renacentista en 'El coloquio de los perros.'" *Actas del V Congreso Internacional de la Asociación de Cervantistas*. Ed. Alicia Villar Lecumberri. Lisbon: Asociación de Cervantistas, 2004. 1559–1572. Print.

———."Zoopoética quijotesca: Cervantes y los Estudios de Animales." *eHumanista/Cervantes* 1 (2012): 448–64. Web. 10 Jan. 2013.

Martín-Estudillo, Luis, Francisco Ocampo, and Nicholas Spadaccini, eds. *Debating Hispanic Studies: Reflections on Our Discipline. Hispanic Issues Online*. University of Minnesota Press. 14. Dec. 2008. Web. 11 Jan. 2012.

Martínez, María Elena. *Genealogical Fictions: Limpeiza de Sangre, Religion, and Gender in Colonial Mexico*. Stanford: Stanford University Press, 2008. Print.

Martínez de Espinar, Alonso. *Arte de ballestería y montería*. Ed. Eduardo Trigo de Yarto. Velázquez: Madrid, 1976. Print.

Marvin, Garry. *Bullfight*. Urban: University of Illinois Press, 1994. Print.

Mastromarino, Mark A. "Teachings Old Dogs New Tricks: The English Mastiff and the Anglo-American Experience." *The Historian* 49.1 (1986): 10–25. Print.

Mateos, Juan. *Origen y dignidad de la caza.* Madrid: Sociedad de Bilióñlos Españoles, 1928. Print.

Mckendrick, Melveena. "Honour/Vengeance in the Spanish "Comedia": A Case of Mimetic Transference?" *The Modern Language Review* 79 (1984): 313–35. Print.

Melville, Elinor G. K. *A Plague of Sheep: Environmental Consequences of the Conquest of Mexico.* Cambridge: Cambridge University Press, 1994. Print.

Messerli, Luisa Rubini. "The Death of the Royal Dwarf." *Beiträge zur kulturwissenschaftlichen Erzählforschung. Hans-Jörg Uther zum 65.* Ed. Rolf Wilhelm Brednich. Berlin: Walter de Gruyter, 2009. 327–52. Print.

Metzger, Richard. "Operation Chihuahua Airlift." *Brand X.* Web. 6 Jan. 2010.

Millares Carlo, Agustín. "La biblioteca de Gonzalo *Argote de Molina,*" *Revista de filología española* 10 (1923): 136–52. Print.

Miramón, Alberto. "Los perros en la conquista." *Boletín Cultural y Bibliográfico* 16.3 (1979): 116–25. Print.

Miramon, Charles de. "Noble Dogs, Noble Blood: The Invention of the Concept of Race in the Late Middle Ages." In Eliav-Feldon, Isaac, and Ziegler 200–16.

Moffitt, John F. "Diego Velázquez, Andrea Alciati and the Surrender of Breda." *Artibus et Historiae* 3.5 (1982): 75–90. Print.

Molho, Mauricio. "Para una lectura psicológica de los cuentecillos de locos del segundo Don Quijote." *Cervantes: Bulletin of the Cervantes Society of America* 11.1 (1991): 87–98. Print.

Molnar, Michael. "Of Dogs and Doggerel." *American Imago* 53.3 (1996): 269–80. Print.

Montaigne, Michel de. *The Essayes of Michael Lord of Montaigne,* vol. 2. Trans. John Florio. London: Oxford University Press, 1906 (1603). Print.

Montoya Ramírez, María Isabel. "La tradición clásica en una obra del s. XVI: *Diálogos de la montería* de Luis Barahanda de Soto." *Ágora: Estudos Clássicos em Debate* 5 (2003): 107–18. Print.

Moreiras, Alberto. *The Exhaustion of Difference.* London: Duke University Press, 2001. Print.

———. "Mules and Snakes: On the New-Baroque Principle of De-Localization." *Ideologies of Hispanism.* Ed. Mabel Moraña. Nashville: Vanderbilt University Press, 2005. 201–29. Print.

Morrison, Toni. *Song of Solomon.* New York: Vintage, 2004. Print.

Myres, Sandra L. *The Ranch in Spanish Texas, 1691–1800.* El Paso: Texas Western Press, 1969. Print.

Nebrija, Antonio de. *Vocabulario de romance en latín.* Ed. Gerald J Macdonald. Madrid: Castalia, 1973. Print.

Nelson, Bradley. *The Persistence of Presence: Emblem and Ritual in Baroque Spain.* Toronto: Toronto University Press, 2010. Print.

Newmyer, Stephen T. *Animals, Rights and Reason in Plutarch and Modern Ethics.* New York: Routledge, 2006. Print.

Nieto Jiménez, Lidio and Manuel Alvar Ezquerra, eds. *Nuevo tesoro lexicográfico del español (s. XIV – 1726)*. 11 vols. Madrid: Arco Libros, 2007. Print.

Nirenberg, David. "Race and the Middle Ages: The Case of Spain and Its Jews." In Greer, Mignolo and Quilligan 71–87. Print.

———. "Was there Race before Modernity? The Example of 'Jewish' Blood in Late Medieval Spain." In Eliav-Feldon, Isaac, and Ziegler 232–64. Print.

Ovid. *Metamorphoses*. The Ovid Collection. University of Virginia E-Text Center. Trans. Anthony S. Kline. Web. 23 July 2010.

Oviedo, Gonzalo Fernández de. *Historia general y natural de las indias*. Ed. Juan Pérez de Tudela Bueso. 4 volumes. Madrid: Atlas, 1959. Print.

Owst, Gerald Robert. *Literature and Pulpit in Medieval England: A Neglected Chapter in the History of English Letters and Of the English People*. New York: Barnes and Noble, 1961. Print.

Oxford English Dictionary. Oxford: Oxford University Press, 2008. Print.

Palma Chaguaceda, Antonio. *El historiador Gonzalo Argote de Molina: Estudio biográfico, bibliográfico y crítico*. Madrid: Consejo Superior de Investigaciones Científicas, 1949. Print.

Papy, Jan. "Lipsius and His Dogs: Humanist Tradition, Iconography and Ruben's Four Philosophers." *Journal of the Warburg and Courtauld Institutes* 62 (1999): 167–98. Print.

Parra Gordillo, Francisco Javier. "El alano español: Presente y futuro." *Scribd.* Web. 17 June 2012.

Pearson, Susan J. and Mary Weismantel. "Does 'the Animal' Exist? Toward a Theory of Social Life with Animals." *Beastly Natures: Animals, Humans, and the Study of History*. Ed. Dorothee Branz. Charlottesville: University of Virginia Press, 2010. Print.

Pérez, Luis. *Del can y el cavallo, y de sus calidades dos animales de gran instincto y sentido, fidelissimos amigos de los hombres, por el protonotario, Luys Perez, Clerigo, vezino de Pertillo*. Valladolid: Adrian Ghemart, 1568. Print.

Pérez, Mirzam C. "Dogs, Demons, and Temptations: The Curious Case of Saint Rose of Lima." American Comparative Literature Association (ACLA) Conference, New Orleans. 2010. Lecture.

Pérez Sánchez, Alfonso Emilio. *Historia del dibujo*. Madrid: Cátedra, 1986. Print.

Pérez de Villagrá, Gaspar. *Gaspar Pérez de Villagrá's Historia de la Nueva México, 1610: A Critical and Annotated Spanish-English*. Eds. Alfredo Rodríguez and Miguel Encinias. Albuquerque: University of New Mexico Press, 1992. Print.

Périers, Bonaventure des. *Cymbalum mundi. Four Very Ancient Joyous and Facetious Poetic Dialogues*. Trans. Bettina L. Knapp. New York: Bookman, 1965. Print.

La pícara Justina. Ed. Julio Puyol y Alonso. Sociedad de Bibliófilos Madrileños: Madrid, 1912. Print.

Picón, Jacinto Octavio. *Vida y obras de Don Diego Velázquez*. Madrid: Librería de Fernando Fe, 1899. Print.

Pidal, Ramón Menéndez. "La Jura de Santa Gadea." *Flor nueva de Romances Viejos*. Madrid: Espasa, 1989. Print.

Plutarch. "Whether Land or Sea Animals Are Cleverer." *Plutarch's Moralia in Fifteen Volumes*, vol. 12. Trans. Harold Cherniss and William C. Helmbold. Cambridge: Harvard University Press, 1958. Print.

Quevedo, Francisco de. "Los enfadosos." *Obra poética, IV*. Ed. José Manuel Blecua. Madrid, Castalia: 1970. Print.

Quirós, Francisco Bernardo de. *Dos comedias burlescas del Siglo de Oro*. Ed. Ignacio Arellano y Carlos Mata Induráin. Reichenberger: Kassel, 2000. Print.

Raber, Karen. "A Horse of a Different Color: Nation and Race in Early Modern Horsemanship Treatises." *The Culture of the Horse: Status, Discipline and Identity in the Early Modern World*. Eds. Karen Raber and Treva J. Tucker. New York: Pargrave MacMillan, 2005. 225–44. Print.

Rahn Philips, Carla and William D. Philips, Jr. *Spain's Golden Fleece: Wool Production and the Wool Trade from the Middle Ages to the Nineteenth Century*. Baltimore: John Hopkins University Press, 1997. Print.

Rangel, Nicolás. *Historia del toreo en Mexico. Epoca colonial. 1629–1821*. Manuel León Sánchez: Mexico, D. F. 1924. Print.

Reutersward, Patrik. "The Dog in the Humanist's Study." *Konsthistorisk Tidskrift* 1981 (50:2): 122–35. Print.

Resina, Juan Ramón. *Del hispanismo a los estudios ibéricos. Una propuesta federativa para el ámbito cultural*. Biblioteca Nueva: Madrid, 2009. Print.

Riley, E. C. "Cervantes, Freud, and Psychoanalytic Narrative Theory." *Modern Language Review* 88 (1993): 1–14. Print.

Río Yenes, Agustín del. *Alano español: historia de una realidad consolidada*. Gijón: A. del Río, 1995. Print.

Rocca, Bartolomeo della (Cocles). *A Brief and Most Pleasant Epitomy of the Whole Art of Phisiognomie*. Trans. T. Hull. London: Waylande, 1556. Print.

Rohman, Carrie. *Stalking the Subject: Modernism and the Animal*. New York: Columbia University Press, 2009. Print.

Rojas, Fernando de. *La Celestina*. Madrid: Alianza Editorial, 1969. Print.

———. *Celestina*. Trans. Mack Hendricks Singleton. Madison: University of Wisconsin Press, 1958. Print.

Rosal, Francisco del. *Diccionario Etimológico. Alfabeto primero de Origen y Etimologia de todos los vocablos originales de la Lengua Castellana*. Ed. Gomez Aguado. Madrid: CSIC, 1992. Print.

———. *La razón de algunos refranes. Alfabetos tercero y cuarto de origen y etymolgía de todos los vocablos de la lengua castellana*. Ed. B. Bussell Thompson. London: Tamesis, 1975. Print.

Rouse, John E. *The Criollo: Spanish Cattle in the Americas*. Norman: University of Oklahoma Press, 1970. Print.

Salisbury, Joyce E. *The Beast Within: Animals in the Middle Ages*, Second Edition. London: Routledge, 2011. Print.

Salomon, Frank. *Ethnic Lords of Quito in the Age of the Incas: The Political Economy of North-Andean Chiefdoms*. Thesis, Cornell University, 1978. Print.

Sanchez, Melissa E. "'Use Me But as Your Spaniel': Feminism, Queer Theory, and Early Modern Sexualities." *PMLA: Publications of the Modern Language Association of America* 127.3 (2012): 493–511.

Santa Cruz, Melchor de. *Floresta española.* Ed. Maximiliano Cabañas. Madrid: Cátedra, 1996. Print.

Schiebinger, Londa. *Nature's Body. Gender in the Making of Modern Science.* Boston: Beacon Press, 1993. Print.

Schiesari, Juliana. *Beasts and Beauties: Animals, Gender, and Domestication in the Italian Renaissance.* Toronto: University of Toronto Press, 2010. Print.

———. "'Bitches and Queens': Pets and Perversion at the Court of France's Henri III." *Renaissance Beasts: Of Animals, Humans, and Other Wonderful Creatures.* Ed. Erica Fudge. Urban: University of Illinois Press, 2004. 70–83. Print.

Schmitt, Jean-Claude. *The Holy Greyhound: Guinefort, Healer of Children since the Thirteenth Century.* Trans. Martin Thom. Cambridge: Cambridge University Press, 1983. Print.

Schwartz Lerner, Lia. "De la 'erudición noticiosa': El motivo de Acteón en la poesía áurea." *Actas del X Congreso de la Asociación Internacional de Hispanistas.* Ed. A. Vilanova. Barcelona: PPU, 1992. 551–61. Print.

Schwartz, Marion. *A History of Dogs in the Early Americas.* New Haven: Yale University Press, 1997. Print.

Seseña, Natacha. "El búcaro de las Meninas." *Velázquez y el arte de su tiempo.* Ed. Enrique Arias Anglés. Madrid: Alpuerto, (1991): 39–48. Print.

Shakespeare, William. *Julius Caesar*: Bartleby.com. Web. 11. Dec. 2008.

———. *King Henry VIII.* Read Book Online. Web. 11. Dec. 2008.

———. *A Midsummer Night's Dream.* Read Print. Web. 11. Dec. 2008.

———. *The Tempest.* The Complete Works of William Shakespeare. Web. 12. Dec. 2008.

———. *Two Gentlemen of Verona.* Bartleby.com. Web. 11. Dec. 2008.

Shannon, Laurie. "Poor, Bare, Forked: Animal Sovereignty, Human Negative Exceptionalism, and the Natural History of King Lear." *Shakespeare Quarterly* 60.2 (2009): 168–96. Print.

———. "Invisible Parts: Animals and the Renaissance Anatomies of Human Exceptionalism." *Animal Encounters.* Eds. Tom Tyler and Manuela S. Rossini. Leiden: Brill, 2009. 137–57. Print.

Shell, Marc. "Animals That Talk; or, Stutter." *A Journal of Feminist Cultural Studies* 15.1 (2004): 84–107. Print.

———. *Stutter.* Cambridge: Harvard University Press, 2005. Print.

———. "The Family Pet." *Representations* 15 (1986): 121–53. Print.

Shesgreen, Sean. *Engravings by Hogarth.* New York: Dover, 1973. Print.

Shubert, Adrian. *Death and Money in the Afternoon: A History of the Spanish Bullfight*. New York: Oxford University Press, 1999. Print.

Skebelund, Aaron. "Breeding Racism: The Imperial Battlefields of the 'German' Shepherd Dog." *Society and Animals* 16 (2008): 354–71. Print.

Smith, Paul Julian. *Amores Perros*. London: British Film Institute, 2003.

Solís, Antonio de. *Varias poesías sagradas y profanas*. Ed. Manuela Sánchez Regueira. Madrid: Consejo Superior de Investigaciones Científicas, 1968. Print.

Steeves, H. Peter. "Lost Dog, or, Levinas Faces the Animal." *Figuring Animals: Essays on Animal Images in Art, Literature, Philosophy and Popular Culture*. Eds. Mary Sanders Pollock and Catherine Rainwater. New York: Palagrave Macmillan, 2005. 21–36. Print.

Steinberg, Leo. "Velázquez' 'Las Meninas,'" *October* 19 (1981): 45–55. Print.

Thomas, Keith. *Man and the Natural World: Changing Attitudes in England, 1500–1800*. New York: Oxford University Press, 1996. Print.

Thurston, Mary Elizabeth. *The Lost History of the Canine Race*. Kansas City: Andrews and McMeel, 1996. Print.

Topsell, Edward. *The Historie of Foure-Footed Beastes*. William Iaggard: London, 1607. The English Experience: Its Record in Early Published Books Published in Facsimile. No. 561. Da Capo Press: New York, 1973. Print.

Torquemada, Antonio de. *Jardin de flores curiosas*. Ed. Giovanni Allegra. Madrid: Castalia, 1982. Print.

Tortorici, Zeb. "'In the Name of the Father and the Mother of All Dogs': Canine Baptisms, Weddings, and Funerals in Bourbon Mexico." In Few and Tortorici, forthcoming.

Tzvetan, Todorov. *The Conquest of America. The Question of the Other*. New York: Vantage, 1984. Print.

Umberger, Emily. "Velázquez and Naturalism II: Interpreting Las Meninas" *Res* 28 (1995): 95–117. Print.

Vahlne, Bo. "Las meninas de Velázquez: observaciones sobre la escenificación de un retrato real." In Marías, 163–69. Print.

Valadez Azúa, Raúl, Blanca Paredes Gudiño, and Bernardo Rodríguez Galicia. "Entierros de perros descubiertos en la antigua ciudad de Tula." *Latin American Antiquity* 10.2 (1999): 180–200. Print.

Valle, Perla. "Aperreamiento: Diccionario de los elementos constitutivos de los glifos." Sup Infor: Editions sur supports informatiques. 2009. Web. 1 Nov. 2009.

Varner, John Grier and Jeannette Johnson Varner. *Dogs of the Conquest*. Norman: University of Oklahoma Press, 1983. Print.

Vega, Lope de. "Si no vieran las mujeres." 1804. Web. 12 Jan. 2012.

Vignolo, Paolo. "Una nación de monstruos: Occidente, los cinocéfalos y las paradojas del lenguaje." *Revista de Estudios Sociales* 27 (2007): 140–49. Print.

Waite, Geoffrey. "Lenin in Las Meninas: An Essay in Historical-Materialist Vision." *History and Theory* 25.3 (1986): 248–85. Print.

Wardropper, Bruce. "La eutrapelia en las Novelas ejemplares de Cervantes." *Actas del séptimo congreso de la Asociación Internacional de Hispanistas*. Ed. Giuseppe Bellini. Roma: Bulzoni, 1982. 153–69. Print.

Wedde, Ian. "Walking the Dog." *Knowing Animals*. Eds. Laurence Simmons and Philip Armstrong. Leiden: Brill, 2007. 266–88. Print.

Werner, Louis. "Dog Tails of the New World." *Americas* 51.5 (1999): 40–47. Print.

Wessem, Haja van. *Cocker Spaniel Inglés*. Barcelona: Editorial Hispano Europea, 1999. Print.

White, David Gordon. *Myths of the Dog-Man*. Chicago: University of Chicago Press, 1991. Print.

Wolchover, Natalie. "Do Animals Know Right from Wrong?" *MSNBC.com*. 2 Nov. 2011. Web. 11 Jan. 2012.

Wolfe, Cary. "Exposures." *Philosophy and Animal Life*. Stanley Cavell, Cora Diamond, John McDowell, Ian Hacking, and Cary Wolfe, eds. New York: Columbia University Press, 2008. 1–42. Print.

———. "Flesh and Finitude: Thinking Animals in (Post)Humanist Philosophy." *SubStance* 37 (2008): 8–36. Print.

———. "Human, All Too Human: 'Animal Studies' and the Humanities." *PMLA: Publications of the Modern Language Association of America* 124 (2009): 564–75. Print.

———, ed. *Zoontologies: The Question of the Animal*. Minneapolis: University of Minnesota Press, 2003. Print.

Woolf, Virginia. *Flush: A Biography*. New York: Harcourt, Brace and Company, 1933. Print.

———. *Mrs. Dalloway*. Hertfordshrine: Wordsworth Classics, 2000. Print.

Žižek, Slavoj. *In Defense of Lost Causes*. London: Verso, 2008. Print.

Index

For Product Safety Concerns and Information please contact our EU representative GPSR@taylorandfrancis.com
Taylor & Francis Verlag GmbH, Kaufingerstraße 24, 80331 München, Germany

www.ingramcontent.com/pod-product-compliance
Ingram Content Group UK Ltd.
Pitfield, Milton Keynes, MK11 3LW, UK
UKHW011018180926
475701UK00004B/816

* 9 7 8 1 1 3 8 2 7 1 9 0 6 *